Higher Education and First-Generation Students

Higher Education and First-Generation Students

Cultivating Community, Voice, and Place for the New Majority

Rashné Rustom Jehangir

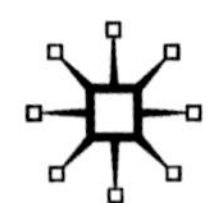

First published in hardcover in 2010 by PALGRAVE MACMILLAN® in the United States—a division of St. Martin's Press LLC, 175 Fifth Avenue, New York, NY 10010.

Where this book is distributed in the UK, Europe and the rest of the world, this is by Palgrave Macmillan, a division of Macmillan Publishers Limited, registered in England, company number 785998, of Houndmills, Basingstoke, Hampshire RG21 6XS.

Palgrave Macmillan is the global academic imprint of the above companies and has companies and representatives throughout the world.

Palgrave® and Macmillan® are registered trademarks in the United States, the United Kingdom, Europe and other countries.

ISBN: 978-1-137-29323-7

Library of Congress Cataloging-in-Publication Data is available from the Library of Congress.

A catalogue record of the book is available from the British Library.

Design by Scribe Inc. (www.scribenet.com)

First PALGRAVE MACMILLAN paperback edition: February 2013

10 9 8 7 6 5 4 3 2 1

This book is dedicated to the memory of my father, Rustom Sorab Jehangir.

The memory of my father is wrapped up in
white paper, like sandwiches taken for a day at work.
Just as a magician takes towers and rabbits
out of his hat, he drew love from his small body,
and the rivers of his hands
overflowed with good deeds

—Yehuda Amichai
(Translated by Azila Talit Reisenberger)

Contents

Acknowledgments ix

Introduction 1

Part I Getting There: First-Generation Students and the Road to College

1 A Long Way from Home: Deepening Our Understanding of First-Generation Students 13

2 Strangers without Codebooks: Isolation and Marginalization 29

Part II Being Here: Surviving the Transition to College

3 Reimagining the University: Theoretical Approaches to Serving First-Generation College Students 47

4 Toward Community, Connectedness, and Care 73

Part III Getting Through: Lessons from First-Generation Students

5 Rationale and Design for the Multicultural Learning Community 95

6 Belonging and Finding Place 119

7 Claiming Self: Identity and Academic Voice 139

8 On Critiques and Possibilities 167

References 187

Index 207

Acknowledgments

When I was about 14 years old, I would often spend my Sunday afternoons with my father, who liked to listen to jazz and talk about the future and the state of the world. He would ask a lot of questions about everything but especially about my hopes and dreams and, of course, how I planned reach these goals. I found these questions alternatively challenging and intriguing. Since that time, I have had the good fortune to work with many other people who have, in similar form, pushed me, nurtured me, and continued to ask questions that challenge and intrigue me. This book and, indeed, my life's work as an educator and teacher have been shaped by these many individuals who have been critical parts of my journey. Working in the General College and the Department of Postsecondary Teaching and Learning at the University of Minnesota has given me a community of colleagues whose engaged teaching, scholarship, and commitment to the idea of education as a public good is at the core of this book.

While so many colleagues have guided me along the way, I would like to make specific mention of a select few: To Bruce and Sharyn Schelske, who have spent 30-plus years dedicated to TRIO programs and first-generation, low-income students—thank you for giving me my first job and for giving me the latitude to explore life as a teacher. Your ability to collaborate and build coalitions was critical to the creation of the Multicultural Learning Community without which this book would never have been. Working with you has deeply shaped the trajectory of my career and reminded me that the fight for equity in education requires tenacity, integrity, and, of course, good humor. To Pat James and Pat Bruch, my co-teachers in our learning community, your creativity, camaraderie, and constant collaboration allowed us to practice "learning community" along with our students. Pat James, your insights, edits, and feedback on many chapters of this book were so important to telling this story and doing justice to our students' voices. Sharing this writing process with you was an extension of our work together in the classroom and reminds me again and again that gifted teachers like you never stop teaching even if when they escape to cabins in the woods.

Jeanne Higbee, thank you for your guidance and advice when I was writing the book proposal and for making the time to listen and support me over these many years. To Bob Poch, thanks for taking the time help me better articulate the core goals of the book in my proposal revision—you are a generous and terrific colleague. To Amy Lee, thank you for your faith in me and for always asking the tough questions—your generous feedback on my earliest drafts helped me envision the direction of this book in my own voice. To Dan Detzner, you have been my mentor, colleague, co-teacher, cheerleader, and constant advisor in this writing process. I am so grateful to you for the prewriting lunches to discuss ideas, for reading my many drafts, and, of course, for taking my calls at all times so that I could process my writing dilemmas. To Barb Hodne, J. T., and Heather Dorsey, my cadre of expert teachers—I appreciated your constant encouragement and the many conversations we have had about the dilemmas and joys of teaching students who are new to the academy.

To my team of amazing graduate students: Jinous Kasravi, Alex Feizter, and Rhiannon Williams—your work conducting the interviews and working with me to analyze, organize, and make meaning of all our students' voices was invaluable. Rhiannon, your five-year tenure working with this project was instrumental to bringing clarity, richness, and depth to our analysis of students' narratives represented in this book. Thanks also for initiating me into more sophisticated electronic data-analysis tools and for finishing my sentences when I got stuck in thought.

I remember also Harvey Carlson and LeRoy Gardner, who deeply shaped my understanding of multicultural teaching and learning. Though you are no longer with us, each of you demonstrated by your practice and your very being what it meant to be committed to education as a path to democracy.

I am indebted to my many students who gave so generously of themselves in my research and in the classroom and entrusted me with their stories, opinions, and perceptions of higher education with candor and grace. You are coauthors of this text, and I hope I do justice to your narratives. I am so grateful to the many students, past and present, who have stayed in touch and taken the time to ask about the progress of the book, reminding me again and again that their stories needed to be told.

To my little family—Mike, my partner in *all* things, thanks for your continual faith in me, for being my in-home research consultant, and for the many Saturday mornings you spent with Riya so that mummy could write. And to my youngest teacher, my daughter Riya, may the classrooms of your future be both challenging and nurturing spaces worthy of your wise spirit.

To my sisters, Nazneen and Nina, thanks for your cheering me on from afar. And for my parents, Aruna and Rustom Jehangir, thanks for the many sacrifices you made so that I could go across the world to "become educated." You have always been my best teachers, and my earliest education at home with you is still the experience that most deeply defines me. While my father died early in my youth, I remember him as an unconventional learner, a creative thinker, and a master storyteller. I am grateful to him for travelling with me in heart and mind as I wrote this book.

The author thanks the following for permission to use text from previous published articles. Springer Press (Jehangir, R. [April 2009]. Cultivating voice: First generation students seek full academic citizenship in multicultural learning communities. *Innovative Higher Education*, *34*[1], 33–49.) and the Pell Institute for the Study of Opportunity in Higher Education (Jehangir, R. [2008]. In their own words: Voices of first generation students in a multicultural learning community. *Opportunity Matters*, *1*[1], 22–31. A publication of the Pell Institute, Washington, DC.).

The author also thanks Azila Talit Reisenberger for her generous permission to use her translation of Yehuda Amichai's poem "My Father" in this book.

Introduction

Cages. Consider a birdcage. If you look closely at just one wire in the cage you cannot see the other wires. If your conception of what is before you is determined by this myopic focus, you could look at that one wire, up and down the length of it, and be unable to see why a bird would not just fly around the wire anytime it wanted to go somewhere. . . . It is only when you step back, stop looking at the wires one by one, microscopically, and take a macroscopic view of that whole cage, that you can see why the bird does not go anywhere; and then you will see it in a moment.

—*Marilyn Frye, "Oppression"*

Jon, a first-generation, low-income college student, described the connections he found between his lived experience and the material he was reading for a course in a first-year college learning community. He said, "The birdcage metaphor is what affected me the most; it was an epiphany. As I was reading it, I recalled a conversation I had with a female friend of mine. We spent hours talking about a problem she had. I really just wanted her to get past it and feel happy. 'You just don't know how it is to be girl,' she said. When I asked her to tell me about what it is like to be a girl, she couldn't properly explain it to me, but now I know that there were many factors affecting who she was."

The birdcage metaphor is a reference to Frye's (2001) article on oppression in which she likens systems that oppress or isolate to a birdcage: to look at a single bar of a cage is to underestimate the problem; rather, a macroscopic view of the entire cage is necessary to see the many facets of the issue at hand. In his effort to understand a friend's dilemma, Jon was also made aware of the many aspects of her identity. My intention is to bring this same lens to bear on our understanding of first-generation students in college. We know a great deal about students who are the first in their family to attend college. Research has focused on precollegiate factors, access to college, enrollment patterns, and financial aid, all of which are important parts of the larger story—but none alone can tell the full story. This book draws upon firsthand narratives of first-generation (FG), low-income (LI) students in order to illuminate how these various bars of the birdcage can limit students. It is also important to consider how the intersections of these limitations may be different for each student, depending on his or her context, social location, and the nature of his or her experiences in educational settings. This book is also a call to reexamine how the academy, particularly large four-year institutions, has underserved LI, FG students and to consider how the changing demographics of this nation demand renewed energy, creativity, and attention to undergraduate education.

FG students are by no means a homogeneous group; rather, their educational journeys represent an intricate juncture of place, aspiration, and access to American higher education. These students have complex identities, making them hard to pigeonhole. They are more often than not students of color, immigrants, and they come from lower socioeconomic backgrounds. They are also parents, employees, and caretakers of their extended families and communities. They all do not have the same story, but aspects of their narratives weave together to form a pattern reflecting both the richness they bring to campuses and the obstacles they encounter in academia. Drawing upon their myriads of voices, experiences, and reflections, this book seeks to demonstrate how their insights interface with what we, as educators, think we know about FG students. What can we learn from these students? How might their insights inform and shape the learning spaces we create for them?

Despite their heterogeneity, what binds FG students together are their aspirations and hopes that a college education will translate into greater opportunities and security for their families and their communities. As newcomers to higher education, they also share the challenge of traversing multiple worlds that appear to have little in common. As a result, they must become adept at wearing many hats and switching from one "disguise" to another in ways that can limit their full engagement in any aspect of their lives. Zahara, an LI, FG student, addressed this struggle between her multiple and competing identities:

> I had class in the afternoon, so I would work for my mom in the morning, then I would come into class. My mom owns a cleaning business. After class, we'd go back to work till like three o'clock or four o'clock. But I had two other jobs and I did school. I was an adult—kind of an adult; I had bills to pay, cell phone, a life to keep up . . . all those other things. I had to do it, because my mom needed me. So it was important for me. Then my mom had a baby girl and I would help feed her, wait for my mom to get home, and then I could go live my life and do homework, go to school, and then I'd do it over again. When I started

> my first semester here, I was on the verge of like shutdown because I was helping my mom with the baby, and I was trying to party and live life and work two jobs. . . . I literally would be on the highway with my little cousin Jason, and I would be like, "Jason, just slap my hand to keep me up." That is how I used be. . . . I used to go that much.

To read Zahara's words is to taste the sense of urgency she feels as she juggles the multiple roles in her life. She is a self-described semiadult, one who bears the many responsibilities of adulthood—bills, childcare, and jobs—while also trying to develop an identity as a college student. Zahara's story typifies the ways in which LI, FG students must navigate many conflicting life roles, not only to gain access to higher education but also to sustain themselves through their academic journey. Her voice is one of many that this book draws upon to contextualize the experience of FG college students by demonstrating the intersections of their varied identities, life roles, expectations, and aspirations. Her story is representative of the many elements that shape the higher-education journey of many historically underrepresented students in college. As Zahara juggles her roles as caretaker, worker, daughter, and student, they all interconnect such that no one role can be evaluated without consideration of the others. Success in one aspect of her life does not translate into success in all; rather, these competing roles often constrain each other. As such, this book is not a work of theory but instead a theoretically informed study of the lived experience of FG students.

For the past 15 years, my work with FG college students has shaped my thinking about education as an idea and a promise for the public good, particularly at large four-year predominantly White institutions. I came to this work through my own journey in American higher education, where my experiences as a foreign student, immigrant, and person of color shaped the lenses through which I lived, worked, and made meaning of the privileges and constraints of postsecondary education. In my work, I have been challenged to make sense of the role that institutions, teachers, staff, and students play in enhancing or diluting the promise of collegial access. I began my formal work in higher education as an advisor in a TRIO student services program, or TRIO Student Support Services (TRIO SSS), a federally funded program situated within a college campus. The TRIO SSS programs across the nation support LI, FG students in their transition to and progress in college by providing a structured first year, intrusive advising, and bridges between academic and student services that facilitate and scaffold students' transition to college.

Later on, I taught supplemental instruction seminars for TRIO students that were linked to challenging first-year courses. What struck me most about this work is what I learned from students—at first it was related specifically to classroom issues, relationships (or the lack thereof) with professors, financial aid, and campus resources. Once we got past these basics, students revealed more and more about the nature of their high school experiences, their perceptions of college, their hopes and fears, and the challenges that confounded their ability to meet their goals as students. These challenges included academic preparation but also involved the many life roles and identities that students juggled. In addition, the students often questioned if they belonged on campus or if they were uninvited guests, here for a short time until they would be found out as "imposters." Many of the students had worked through high school and often held at least one job in college. Their schedules were jammed with classes, work, and responsibilities at home. They spoke multiple languages while carrying scars, both figuratively and literally, from refugee camps and prior educational experiences alike. I was always struck by their resilience and the skills they had at their fingertips. Yet within the educational milieu, these skills were frequently recast as deficiencies, often causing these students to feel like interlopers.

These experiences propelled me to think more about how learning spaces could challenge the isolation and marginalization that FG students experience, both on campus and in the classroom. How might we support FG students and draw upon their multiple identities and lived experience as strengths to engage them in knowledge construction and create a sense of belonging in academia? This question led to the creation of the Multicultural Learning Community (MLC) for FG students in the TRIO program. Together with two other faculty members, I helped design a learning community that became the basis of an eight-year study about the experience of FG students in their first year and beyond. Student voices featured in this book are drawn from the data in this study, both in the form of narrative writing and retrospective interviews.

THE STUDY: CONTEXT

This book relies on data from a study that began in fall 2001 and involves LI, FG students who participated in an MLC during their first year of college at a large, predominantly White public research university. Significant research points to the fact that the lives of these students are often compartmentalized into distinct categories: home world, school world, and social world, to name a few (London, 1989, 1992; Rendón, 1992). As Zahara's story suggests, these disconnections create a feeling of living in divided worlds and a sense of isolation and marginalization on campus.

The intention behind the MLC design was to build bridges between these disparate worlds and invite FG students to find a place of "full citizenship" in the academy. With attention to the multicultural curriculum and critical pedagogy within the interdisciplinary learning community, the MLC was designed to create academic and social integration for these students during their first year of college. The learning community was comprised of three first-year courses: a freshman composition course, a creative arts course, and a course in either the social sciences or literature that examined issues of race, class, gender, and inequity. The curriculum for all three courses focused on the themes of identity, community, and agency. These particular themes were chosen because they invited students to share their stories, cultivate their voices, and grapple with the intersection of academic content and lived experience.

Between fall 2001 and spring 2007, the MLC was offered seven times to seven different cohorts of first-year, LI, FG students. Concurrent enrollment in each learning-community class was required and was capped at 24 students. This curricular linkage allowed the students to examine relationships among various media of expression and to apply diverse interdisciplinary perspectives to their lived experiences, while simultaneously being empowered to consider their own narratives as integral to the knowledge construction process. During the course of each semester, students engaged in weekly reflective writing about the nature of their learning experience and its connection to their lives. This data was the basis of the first part of my study and sought to capture students' perceptions of their learning during enrollment in the MLC.

The second part of the study sought to create a longitudinal view of the college trajectory of students who participated in the MLC. Building on the narratives of students, which captured their first-year transition to college, the study explored if and how the MLC experience challenged the isolation and marginalization of FG students beyond their first year, and if first-year participation had any long-term impact on the rest of their university experience. Students from four different MLC cohorts between fall 2001 and fall 2005 were contacted via mail and email and were invited to participate in the interview. Twenty-four students responded to the invitation and were interviewed when they were juniors, seniors, or recent graduates. Interview questions covered four specific areas: students' MLC experiences, their university experiences outside of the MLC, their involvement in extracurricular activities, and their future goals. Details of the study and the student participants are included in the final section of this introduction.

Students who were interviewed not only reflected on their first year of college but also shared the challenges, successes, and strategies they employed in order to survive in their college journey. Students' voices—both self-reflective and critical—revealed lessons for educators about the complexity of their journey to and through college. They also shed light on the experiences and environments that sustained and engaged them, exposing areas where the academy is repeating the same mistakes again and again.

Drawing upon these student voices, this book argues that LI, FG students experience a heightened sense of isolation and marginalization in college. This separation extends into the mainstream curriculum, pedagogy, and campus life, particularly for FG students whose learning styles and cultural capital are undermined by or left out of the educational experience. If education is indeed a public good, where is there a place for FG students, and what can be done to engage them and include them so they will stay the course toward a degree? How can their lived experience be invited into the knowledge construction process so that all students can engage in the practice of learning in democratic, inclusive communities?

To explore these questions, the book is divided into three parts. The first part is titled "Getting There: First-Generation Students and the Road to College." This part investigates and questions the oftentimes one-dimensional understanding of FG students by exploring the multifaceted experiences of these students. I will address the way that various literature has defined the term "first-generation" and will provide a rationale for my use of this term in the book. This part also attends to the rich context of factors (life roles, demographics, cultural affiliations, academic histories, and self-perceptions) that shape the first year of college for FG students. Each of these factors intersects with higher education in ways that can heighten the isolation and marginalization of FG students on campus. FG, LI students arrive on campus unaware of the explicit and implicit expectations, the unwritten rules, and the historic social memory required to navigate college. Unlike their traditional counterparts, they do not have the "codebook" and must discover the modus operandi in a learning environment that often undermines or ignores their lived experiences, both in and outside of the classroom. By referencing prior research and my own longitudinal study of FG students, this investigation explores the ways in which isolation and marginalization are manifested in the FG experience. This part also foregrounds the interdependence of students' academic and social experiences and self-confidence.

The second part is titled "Being Here: Surviving the Transition to College." This part further explores the isolation and marginalization of FG students in the academy from various theoretical frameworks and examines the importance of focusing attention (as teachers, administrators, and scholars) on understanding and countering those key dimensions of FG students' first-year experience. With attention to the elements of design that can transform academic spaces, I will examine critical pedagogy, an integrated multicultural curriculum, and learning-community design as praxis for empowering students to claim their place in the academy. This part addresses how the confluence of these ways of approaching the classroom that can shape a space that invites students and their lived experience into knowledge construction, creating a sense of belonging on campus. Critical pedagogy and integrated multiculturalism are presented as theoretical responses to the marginalization of students in a historic and social context. Learning-community design and organizational partnerships are vehicles that operationalize critical multicultural pedagogy to help students with complex realities cultivate their intrapersonal, interpersonal, and cognitive development in the academy. This part also employs student voices in order to substantiate the theoretical arguments presented previously and to demonstrate ways in which student perspectives challenge current practices in classrooms and campuses, illustrating how these gaps impact student learning.

The third and final part of the book is titled "Getting Through: Lessons from First-Generation Students." This part examines theory in practice by demonstrating the application of key components from current pedagogies in a contemporary curricular initiative: the MLC. Drawing upon the extended experience of a handful of students, this part employs narrative research in the form of storytelling to demonstrate how the MLC cultivated a sense of place and voice for FG students and to examine the specific curricular and pedagogical strategies that support these outcomes. In addition to discussing the design and rationale for the MLC, the themes from students' narratives are presented as evidence of one model for utilizing critical pedagogy and multicultural curriculum using learning-community design and partnerships between academic and student services. The book concludes with students' critiques of their experience in the academy and invites reflective scholars and practitioners to grapple with the important questions that these critiques raise for the future of our work together.

I continue to believe that despite its problems, American higher education is the most promising practice of democracy in which we can collectively engage. Yet it is a broken promise for many FG students. My hope is that their voices, stories, and insights will speak to the gaps that need our urgent attention if we are to repair and reimagine the academy to include this new majority of students.

NOTES ON RESEARCH METHODOLOGY AND STUDENT PARTICIPANTS

My reflections on the practice of teaching and learning with first-generation college students have come from two sources. The first source comprised personal documents, as represented by the students' weekly reflective writings and their final reflective papers in the MLC. One hundred and thirty students who participated in the MLC in seven separate cohorts between 2001 and 2007 completed these reflections. This portion of the study used an interpretive multiple case study, where, in each of the seven cohorts, each cohort represented one case. This approach captured the students' perceptions of their learning experience *in process* (i.e., as students responded to weekly reflective writing prompts and opened-ended reflections about their learning).

The justification for using personal documents as a data source is not unlike that for observational techniques. Where the latter seeks to capture overt behavior, the former seeks to uncover the "inner experiences: to reveal to the social scientist life as it is lived without the interference of research" (Cook, 1959, p. 325, as cited in Merriam, 1998, p. 116). Critical pedagogues and critical feminists also support the merits of reflective writing as a data source because this type of writing "allows students to define issues, express feelings, and develop descriptive texts for analysis" (Sleeter, 1995, p. 429). The goal of collecting weekly reflective writing was to portray students' perceptions of their experience as it was happening, rather than only retrospectively at the end of the semester. It also created the opportunity for students and teachers to engage in a private dialogue with one another. Students shared reactions, insights, and feelings about the curriculum, peers, teaching, and their learning in increasingly candid ways, which allowed teachers to respond or simply to know and consider what individual students were grappling with in their learning experience and in their lives.

Every week during the semester, students completed these self-directed reflections and were also asked to respond to several questions from weekly "learning logs." These learning logs were adapted from the work of Stephen Brookfield (1995) and sought to guide students' reflections on aspects of their learning through the MLC and to illustrate how their experience was impacted by course content, peers, instructors, and the relationship among the three MLC courses. At the end of the semester, each student was asked to reflect on each of his or her learning logs and become interpretive researchers of his or her own experience. The final assignment asked students to reflect back on their own learning logs and draw out themes and highlights

in their experience, as expressed in their own writing. All s
writing was coded using the process of categorical aggregation, which yielded thematic categories (Lincoln & Guba, 1985, p. 350). Data were then tabulated under the themes and subthemes and were cross-referenced by student race, gender, a numerical student code, and week in the semester. The data were coded by two readers to ensure interpreter reliability or "investigator triangulation" (Stake, 1995).

The second set of data came from 24 semistructured interviews with students who were past participants in the MLC. The research methodology for the interviews was informed by the constructivist paradigm, using narrative and case study data analysis procedures. The interview questions addressed four specific areas: students' MLC experiences, their university experiences outside of the MLC, involvement in extracurricular activities, and their future goals. Overall, we sought to gain a rich and deep understanding of how students thought the MLC experience had affected their university experience. The interviews were digitally recorded and transcribed verbatim. In the first phase of the analysis, I worked with two graduate students to identify emerging themes across eight randomly selected interviews that served as individual cases. In a process of cross-checking, we created categories from these cases and compared them with all the participants' interview texts (Flick, 1998). This process continued until there was a "saturation of categories" (Lincoln & Guba, 1985) that resulted in 10 themes. Of the 10 themes that emerged, four—disequilibrium, claiming the self, academic identity, and critiques—were most salient to student development. To deepen our understanding of the role that curricular experience played in developing personhood, we engaged in a second phase of analysis. In phase two, our team applied Baxter Magolda (2001) and Torres and Hernandez's (2007) theoretical frameworks of self-authorship to the four salient themes in order to examine the intrapersonal, interpersonal, and cognitive development of the LI, FG students who had participated in the MLC.

THE STUDENT VOICES

All students who enrolled in the MLC were participants in the TRIO SSS program, a federally funded nationwide program that supports LI, FG students through college. The particular TRIO SSS program that served the student participants has been funded since 1976, with a mission to serve students who are LI, FG students and students with disabilities. At least two-thirds of the TRIO SSS participants must be first-generation and low-income. The SSS program is described a "multidimensional program that provides academic support, supplemental study groups, learning communities, and leadership development" (Council for Opportunity in Education, 2004) for FG students who seek admission in this program.

Students in the MLC

Over the course of seven cohorts (between 2000 and 2007), 130 students participated in the MLC. Across all seven cohorts, race broke down across cohorts as follows: 44.6 percent Black; 28.4 percent Asian American; 10.7 percent White; 8.4 percent Hispanic; 5.3 percent Native American; and 3 students listing no racial data. Females outnumbered males overall by almost two to one: 84 females (65 percent) to 44 males (34 percent). It is important to note that these racial descriptors were based on university demographic data that students selected, but it does not acknowledge the ethnic or cultural diversity within racial groups. For example, of the 28 percent of Asian American students, 71 percent were first- or second-generation immigrants of Hmong or Vietnamese descent. Of the 44 percent of Black students, 20 percent were students who were first- or second-generation immigrants from the African subcontinent representing Ethiopia, Somalia, Eritrea, and Liberia, and the remaining 24 percent were African American. With regard to demographics, I have chosen to differentiate between the terms *Black* and *African African* to capture the range of identities within this domain, particularly since many African immigrants do not identify themselves as African American.

Students Involved in Retrospective Interviews

Students from four different MLC cohorts between fall 2001 and fall 2005 were contacted via mail and email and were invited to participate in the interviews. The final sample included 24 students of varying races and gender. Racially, the sample was primarily non-White, with students identifying as White (five); Black or African American (fourteen), out of which four were of East African descent; Asian (three); Latina (one); and biracial (two). Forty percent were male and 60 percent were female. At the time of the interviews, 23 of the interviewees were still enrolled at the university or had recently graduated. Only one interviewee had dropped out of the university, which is acknowledged as a shortcoming in this study.

In narrative writing and interviews, students' candid reflections, observations, and critiques of their experience in college have deeply informed my own thinking, writing, and teaching. Their voices have been the central inspiration for this book and carry a sense of urgency that demands to be heard.

Part 1

Getting There

First-Generation Students and the Road to College

FGS in the new universities
(primarily UG)

Chapter 1

A Long Way from Home

Deepening Our Understanding of First-Generation Students

I have to be the first one in my family to graduate from college; I am seen as the golden child. My mother has high expectations of me, and I never want to let her down in any way. See, where I'm from, there are not many opportunities, so when I was given a scholarship to attend the university, I was given the opportunity of a lifetime. It is very important to me that I succeed at this college because I want to be able to get that good paying job, so that my mom won't have to struggle anymore. I want to be able to set the bar and show people that it isn't impossible for an inner city kid who came from a low-class family to be able to graduate from a four-year college. I want to be the one who changes the normal flow of things and shake things up a little. I want to be able to help change the lives of people who are just like me who came from nothing, but is working hard to become something great. With that said, college for me is the lonely way.

—Rhian, African American female student

Rhian's story speaks to the promise and burden of the educational opportunity that many first-generation (FG) students grapple with when preparing to enter college. The journey to college is influenced by many factors that will be addressed in this chapter, but even for those who capture the prize of admission, the reward is weighty in that it carries expectations not only for the student but also for his or her entire family and community. As Rhian's states, her experience in college may be lonely, but her dreams and expectations are crowded with the faces of her mother, her inner-city community, and the demands of doing something that will "change the normal flow of things" in her life. Like Rhian, many FG students and their families want and need the college promise to translate into opportunities that will allow them to collectively cross the boundaries of class, race, and geography into a place of greater economic stability.

The term "first generation," in and of itself, does not have one set definition in academic discourse. Its variance in definition reflects the heterogeneity of this diverse group of students, as well as the extent to which different authors lay claim to particular aspects of the FG experience as central or critical. The very act of naming and categorizing a group is shaped by what one considers to be salient to their experience, their relative location to power (Bruch, Jehangir, Lundell, Higbee, & Miksch, 2005) in the context of resources both implicit and explicit, and how these factors might shape or constrain their educational opportunities. Often, terms created to name or categorize specific groups are used in the literature but can mean different things to different audiences. We most commonly refer to FG students as those who are first in their immediate family to attend college. Billson and Terry's (1982) definition of FG students includes those whose parents have no college experience. Federal TRIO programs charged with creating educational opportunity for disadvantaged students via outreach and college support use a more generous definition of FG status to include students for whom "neither parent has earned a four-year college degree" (U.S. Department of Education: Federal TRIO programs, 2010a). By contrast, the National Center for Educational Statistics (NCES) limits FG status to "students who are first in their family to pursue education beyond high school" (NCES, 2007a). Other researchers suggest that FG status is based on parental education and note that students whose siblings have attended college are still considered first generation (London, 1989; Longwell-Grice & Longwell-Grice, 2007; Nunez & Cuccaro-Alamin, 1998). York-Anderson and Bowman (1991) limit their definition to include students whose parents and siblings have one year or less of college education.

Demographically, FG students are more likely than their more advantaged peers to be students of color, older than 24 years, female, nonnative speakers of English, and born outside the United States. They are also more likely to have a disability, care for dependent children, and be single parents. Compared with their traditional counterparts, they are also more likely to have earned a high school equivalency diploma and be financially independent of their parents (Choy, 2002; Engle & Tinto, 2008).

Income status is another central aspect of the FG experience that impacts precollegiate educational experiences, access, and persistence in college in ways that perpetuate economic inequities between the haves and have-nots in America. Family income impacts not only graduation from high school, a precondition to college access but also impacts enrollment, persistence, and completion of collegiate degrees. While high school completion rates of low-income (LI) students rose from 26 percent in 1972 to 54 percent in 2005, they still pale in comparison to the 82 percent high school completion rate of high-income students (NCES, 2007). Similar trends are evident in college completion rates, where in 2007, bachelor's degree completion rates for students from the lowest income quartile were 24.5 percent, compared with 47.6 percent and 94.6 percent for students in the third and top quartiles, respectively (Mortenson, 2008).

While income is critical to educational mobility, it is not only income per se that determines status but how income intersects with social perceptions of class. As such, other terms connected to the FG experience include *working class* and *blue collar* (Longwell-Grice & Longwell-Grice, 2007; Vander Putten, 2001). These terms seek to demonstrate that income and social class are not synonymous; rather, social class is a "combination of economic status, values, beliefs, and assumptions" (Longwell-Grice & Longwell-Grice, 2007, p. 410). Thus while income may be a numerical determinant of one's financial worth, social class captures the nature of one's life experiences, aspirations, and family expectations, said or unsaid, as shaped by income, work, domicile, and family history. The term *nontraditional* is also often associated with the FG experience, though it does so using broader brush strokes that speak to the relationship between experiences, expectations, and one's aspirations. Rendón (1998) eloquently captures this interplay when she says of nontraditional students, "I am talking about single mothers and fathers coming back to school after ten years, students who have been told that they will never amount to anything, students who have lived in poverty and who are the first in their family to attend college" (p. 2). Embedded in each of these definitions, though, are the ways in which FG students inhabit spaces where the intersection of race, class, and gender impact not only access to college but also their aspirations about their place in the unfamiliar land of higher education. For the purpose of this book, the use of the term *first-generation student* will apply to LI students whose parents do not have bachelor's degrees. Low-income status is based on Federal TRIO programs' annual low-income levels; in 2009, low-income status was $33,075 annual income for a family of four (U.S. Department of Education, 2010b). If the student was raised in a single-parent household, then FG status is based on the educational level of the parent who is the primary caregiver.

THE CONFLUENCE OF RACE AND CLASS

Central to the discussion of FG students and income is the historic relationship between race, class, and upward mobility in the United States (Cabrera & La Nasa, 2000; Rendón, 1992). Ranging from home ownership to employment to educational opportunities, people of color have consistently lagged behind Whites with regard to achieving these milestones in pursuit of the American dream. The relationship between race and class with regard to access, persistence, and completion of college is no different. We know that there are more than 12 million children in the United States living in poverty, two-thirds of which come from minority backgrounds (Engle & Tinto, 2008, p. 5; U.S. Census Bureau, 2003). The fastest growing segments of our population are then LI students of color (Kelly, 2005; Mortenson, 2006). Students of color comprised 33 percent of our high school graduates in 2002, and this number will grow to constitute 40 percent by 2012. Yet despite this growth in number, students of color who are low income, and many who are the first in their family to attend college, are least likely to earn college degrees (Kelly, 2005). While access to college for LI, FG students may be improving, only 45 percent of Pell grant recipients enrolled in four-year colleges in 2001. This number is down from 62 percent in 1973 to 1974 (Engstrom & Tinto, 2008; Mortenson 2003). This 28 percent drop in four-year college enrollment demonstrates yet another way in which stratification of opportunity creates widening gaps with implications for economic prosperity affecting families, communities, and the country as a whole. While students are constrained by race and class, they are also constrained by the ways in which these demographics manifest themselves into systems of racism and classism within and outside of their educational environments. Similarly, the isolation of FG students can be examined more in detail if one sees that the status of being first generation is linked closely to the limitations of racism, classism, and other manifestations of inequity that occur before, during, and after college.

ACCESS TO COLLEGE: WHO GOES WHERE AND HOW LONG DO THEY STAY?

Approximately 24 percent of all students enrolled in postsecondary education are FG and LI (Engle & Tinto, 2008). Drawing on data from the National Center for Educational Statistics, Engle and Tinto (2008; note that roughly 75 percent of these students begin their higher education journey in two-year and for-profit institutions. At public and private four-year institutions, they make up approximately 18 percent and 16 percent of the student body, respectively, where by comparison, their more advantaged peers make up 54 percent of the population. These numbers demonstrate that while FG students are growing in higher education, they are disproportionately represented in two-year programs (McPherson & Shapiro, 1998). In addition, this concentration of FG students in two-year institutions includes a concentration of students from low-income and lower-middle income families, while upper-middle and high-income families tend to be enrolled in four-year colleges. Even at four-year institutions, academic selectivity translates into the most affluent students being served at highly selective four-year colleges and universities, and students from the lowest-income families being represented primarily at four-year open door institutions (Mortenson, 2003). Across all institutional types, LI, FG students were nearly four times as likely to drop out of college after their first year, compared with their peers who did not have these disadvantages. The dropout rate for LI, FG students at two-year institutions was higher than in public or private four-year institutions. Transfer rates from two- to four-year institutions was also impacted by LI, FG status, where among students who began their studies at "public two-year institutions or for-profit institutions, 74 percent of LI, FG students did not transfer anywhere within six years compared to 38 percent of their most advantaged peers" (Engle & Tinto, 2008, p. 13). This trend substantiates previous research that enrollment in a two-year program substantially reduces one's chance of eventually earning a four-year degree (Breneman & Nelson, 1981; Karabel, 1986; Tinto, 1987). In an effort to give meaning to the uncertain road to higher education for LI, FG students, precollegiate factors must be considered, as well as those that impact transition to and persistence in college. These issues are certainly not mutually exclusive, but each may become more pressing at different junctions of a student's journey to and in college.

A Crushing Debt: Paying for College

I looked at all the pictures and everybody in there, you know; there were some people that I don't even see on this campus. They are not continuing in school, and I used to get mad, and I used to say, how come you are dropping out of freshman year? You know, that is one thing that I used to say: how do you drop out of freshman year? As I learned in the classroom and stuff, I started understanding; money . . . this university is expensive. That is the number one thing that I have learned, freshman year it's easy, I am not going to lie, but it's not easy on your pocketbooks. It's expensive, and that's why people drop out as I see it.

—Davu, African immigrant male student

Davu's narrative reflects how education is still viewed as an opportunity to break away from poverty but also how financial support is a critical step to converting the promise of college admission into a reality. First, it is important to note that while the costs of college have increased consistently, median family income has remained fairly stagnant (Mortenson, 2008). Engle and Tinto's (2008) study of three different data sets from the National Center for Educational Statistics, ranging from 1992 to 2004, demonstrates that LI, FG students pay less to attend college than their higher-income counterparts, but this is because they are more likely to attend lower-cost institutions, such as two-year colleges. Despite making enrollment choices that are fiscally conservative, LI, FG students have high financial need and receive "only slightly more financial aid than their peers resulting in a shortfall of approximately $3,600 even after loans are accounted for" (Engle & Tinto, 2008, p. 22). If these students were to forgo loans, their average out-of-pocket expense would be close to $6,000. LI, FG students are often working more than one job, not only to support themselves and their families but also to pay for educational expenses. With most holding part- or full-time jobs while attending college, Archer and Lamnin (1985) found that FG students who were also ethnic minorities expressed concern over financial pressures more often than White FG students. The trend toward financial aid policy that has perpetuated a two-tiered, class-based educational system is shaped by sharp reductions in state support of higher education, which in turn has resulted in higher tuition rates. The domino effect continues as public and private institutions shift their selection process toward more affluent students. While they may admit lower-income students, the financial aid packages provided tend to be merit versus need based, thereby limiting any real possibility of matriculation for LI, FG students. Finally, federal financial aid based on need, such as Pell grants, has not grown to match the needs of LI, FG students; rather, these students are more likely to be offered nonsubsidized loans and tax credits (Mortenson, 2003; St. John, 2002, 2005). The Pell Grant program is the single largest source of need-based college funding in the United States, but the value of this grant has fallen while the cost of college has increased. In 1980, the Pell Grant covered 77 percent of the cost of attending a four-year institution; today, it covers only 36 percent of this cost (Cook & King, 2007).

While grants, scholarships, work-study opportunities, and even loans can help LI, FG students matriculate and persist in college, the accumulation of high debts as a result of loans can impact persistence. For some students, the prospect of incurring high debt actually decreases persistence because they choose to work more hours to pay for school rather than take on more debt. In addition, incurring debt, even in the form of subsidized loans, can be culturally incongruent with some students and their home worlds (Bowen, Chingos, & McPherson, 2009). Finally, as loans constitute a more significant portion of the financial aid packages for LI, FG students, the "cumulative loan debt" (Engle & Tinto, 2008, p. 23) incurred by students who leave higher education before completing their degrees is staggering. For example, LI, FG students who left public and private four-year degrees after one year of enrollment owed an average of $6,557, while those who left in the fourth year were saddled with $16,548 of debt. The 2009 federal poverty income guidelines (Federal Poverty Income Guidelines, 2009) lists an income of $22,041 as the gross annual earnings for a family of four living in poverty—to compare this income with the potential debt that an LI, FG student can incur without earning a degree demonstrates how students who leave postsecondary education are further disadvantaged. The way financial challenges manifest themselves may vary for FG students, but there are multiple ways in which finances, and lack thereof, influence students' choice of school, matriculation, and persistence.

Academic Histories

> *I remember being nervous as hell. Just that, personally I took . . . I took a five-year break from high school to college. I was 23 when I was going to college. Most of my academic experience up to that point wasn't really good. So going to college was like, "What the hell am I doing here?" Parents didn't go to college. So I didn't have any background of what the hell college was. Really not understanding how academia works. And, you know, growing up and being in high school, I didn't have a great experience at all. I was really kind of made to feel like my ideas weren't good, and that was part of the reason for my nervousness.*
>
> —*Jarod, White male student*

Jarod's words characterize the way in which many FG students feel about school and the process of understanding the inner workings of school systems and teacher expectations. Like most students, the learning experiences of FG students are colored by the type of secondary experience they had. Unlike other students, though, their demographics impact school choice and options, and these precollegiate experiences are typified by secondary schools with limited resources, curricula, and extracurricular activities. In other cases, as Jarod explains, the feeling of being marginalized starts even before college, and FG students see themselves as outsiders in the educational context even before they arrive to college.

With regard to high-school location and type, FG students are more likely to have attended a high school in a small town or rural area and are slightly more likely to have attended a high school where more than three-fourths of the student body is comprised of under-represented students of color (Warburton, Bugarin, & Nunez, 2001). Finally, 92 percent of FG college students attended public schools (Warburton et al., 2001).

LI, FG students are also less likely to have access to rigorous high school curricula and advanced placement courses, and they tend to have limited experience with time-management and study skills. One LI, FG student of color reflected back on her high school experience and reported, "It seemed like we had gaps. It was like we were missing part of the picture. . . . I kept thinking, 'Gosh, did I miss some lectures along the way?' But I didn't; they just weren't there," (Richardson & Skinner, 1992, p. 32). For some LI, FG student, these gaps translate into anxieties and lack of confidence about their ability to manage academic content in college and to navigate the bureaucratic idiosyncrasies of campus life (Bui, 2002; Cabrera, Terenzini, & Bernal, 1999; Chen & Carroll, 2005; Nunez & Cuccaro-Alamin, 1998; Richardson & Skinner, 1992). More specifically, once in college, FG students tend to be overrepresented in remedial or developmental courses when compared to their more advantaged peers (Engle & Tinto, 2008). Adelman's (1999) study on academic intensity and attendance patterns on bachelors' degree attainment found that the chances of earning a four-year college degree were deeply impacted by the access to academic resources and the "intensity and quality of secondary school curriculum" (p. 2). Specifically, academic resources included a composite of high school curriculum, test scores, and class rank. In context of these measures, students from the lowest socioeconomic standing were less prepared in comparison to their peers from high socioeconomic backgrounds. Yet despite these tough odds, students from the lowest socioeconomic standing who had the highest access to academic resources completed four-year degrees at a rate exceeding that of the majority of their high-income peers, suggesting that socioeconomic status need not negatively influence students for the duration of their college career.

ROLE OF PARENTS AND COMMUNITY

When I first came to the U . . . I didn't really have any expectations. I was just lost, like I said I only applied to one school. So I was just like, "OK, if I am going to college, I go, and if I don't, I don't. . . ." Before my senior year of high school, I was a manager at a packaging company, and everybody who works at the packaging company for the most part was a college student because they provide so much college tuition. So I spoke with them a lot, nobody that I really associated with went to the U at the time. And I remember them talking to me about the programs, the games, the parties, just all kinds of activities and things, that was the big superficial reason. But what you do know from your past 12 years, is the academic aspect, and it's a big question mark and you're not sure you know how hard it is going to be, if you're going to meet people, if people are going to like you. One of my favorite movies growing up was the movie called Higher Learning. And that was always my perspective of college. I didn't know I was a high school kid, and I didn't really have too many people around that even went to college. So I kind of had mixed views about just going to college.

—Tyrone, African American male student

As Tyrone's comments suggest, the road to college is an uncertain one for FG students, and for some, applying to college may be an afterthought or happenstance simply because the process of applying, attending, or aspiring to college is an unknown experience for them and their families. While parents' educational level is a strong determinant of college matriculation, parents' perceived expectations, support, or values about attending college also impacts the journey toward postsecondary education for FG students.

The journey to college can be thought of in stages: predisposition, preparation, and matriculation. Predisposition refers to the extent to which family background, beliefs, and values concerning education predispose students toward a certain view of college (Hossler & Gallagher, 1987; Hurtado, Inkelas, Briggs, & Rhee, 1997). Richardson and Skinner (1992) use the term *opportunity orientation* to capture the "beliefs students develop about valued adult roles and about the part played by education in structuring access to these roles" (p. 30). It is not surprising, then, that encouragement to attend college, as well as saving for college and talking about it as an expectation, influences postsecondary educational plans for students as early as ninth grade (Conklin & Dailey, 1981; Murphy, 1981; Stage & Hossler, 1989).

Generally, second-generation college students' parents are familiar with and prepared to coach them through the process of applying and acclimatizing to the expectations of college. FG students, however, have a very different experience with their families. It is here that the second stage of preparation in the college process comes into play. The process of preparing to matriculate and choose a college involves parental expectations, support, and understanding of the college and financial aid application process. It can also involve other sources of social capital that shape one's life trajectory. Social capital is defined as "networks of people and community resources" (Yosso, 2005, p. 79). Parents who had firsthand knowledge of college are often better equipped to inform, guide, and shepherd their children through the process of making choices and applying to different schools. On the other hand, parents of FG students are less likely to have the know-how to navigate their children through the complex processes of college applications and choice, and they may also have a limited understanding of how to acquire financial resources for college (Choy, 2000; McDonough, 1997). FG parents may also struggle between wanting their children to work to contribute to the family income while simultaneously seeing that a college degree could bring them greater economic stability (Terenzini, Rendón et al., 1994). In some cases, students may attend community college because these institutions appear to be the most direct link to developing particular skills and entering the workforce to procure a secure job (Philippe & Valiga, 2000). To consider how LI, FG students make these choices is also to consider the cultural and filial frameworks from which they derive meaning. Research on Latino/a college choice demonstrates that these students greatly rely not only on their parents but also on members of extended family and peers to assist in choice and information gathering about college (Pérez & McDonough, 2008). Access to a strong social capital network can enhance and motivate students toward postsecondary educational opportunities despite other constraints, and for some, proximity to family may facilitate success in college. The challenge is to discern between strong and weak social capital networks, as not all students have access to the latter, which may have negative repercussions for their postsecondary journey. As one female student expressed, "My mother makes me go wherever my sister goes, but that is not my real choice" (Pérez & McDonough, 2008, p. 257).

Beyond parents' stated expectations, there are socioemotional repercussions of parents' perceived support, especially for LI, FG students who are in the third stage of matriculation to college. Billson and Terry's (1982) study of the effects of parental education on the attrition of FG and second-generation college students found that although FG parents were emotionally supportive, second-generation students received both emotional and financial support from their parents. York-Anderson and Bowman's study (1991) focused on the extent to which FG and second-generation students differ in knowledge involving college and perceived support from their family. Consistent with Billson and Terry (1982), their findings reveal that FG students felt less family support than their more advantaged peers. They conjecture that this lack of support is primarily because FG parents know so little about college themselves, and that despite intentions to be supportive, these parents do not have the resources or insights about college to offer support that could be useful. Howard (2001), a college professor who began his academic life as an LI, FG student reflected on the parental role in his own educational endeavors: "My parents wanted me to be successful in school; they just didn't know how to help me be successful" (p. 11). There is also a complex push-and-pull aspect embedded in how FG parents might express their support. There may be some mixed messages that further complicate the experience of FG students in college. Parents might expect good grades and success but not at the cost of reducing commitment to family roles or contributions to the family income. Often it is that the unsaid expectations are even harder for FG students to negotiate: in other words, "go out and learn things, be smart, but not too smart." Leaving home brings inevitable changes for all students, and these change are coupled with the developmental changes of emerging adulthood. Yet for the parents and families of FG students, their child might begin to express ideas, aspirations, and perceptions of the world that are both alien and contradictory to their own values (London, 1989; Rendón, 1992). This can create a feeling of losing their child anew. Reflecting on her experience as a working-class academic, Carole Leste Law (1995) writes about the push-and-pull within her own family and her life at college:

> At home I could never get myself to talk about books or ideas that never intersected with the lives of my mother, brother, and cousins and extended family. To talk about my studies seemed ridiculous and stuck up at best in a context that seemed as mistrustful of academia as academia was condescending of it. No one in my family ever wrote a college paper, no one ever tried to enroll in classes that were closed, no one put together a degree program, and more to the point, no one ever cared about such things. So you better believe I kept my mouth shut there too. (p. 5)

For LI, FG students, this struggle with family and community might mirror and exacerbate internal struggles they have about their own place in the world and between their home and school worlds (Lara, 1992; London, 1989, 1992; Rendón 1992; Richardson & Skinner,

1992; Rodriguez, 1982; Terenzini, Rendón et al., 1994, Terenzini, Springer, Yaeger, Pascarella, & Nora, 1996).

LIFE ROLES AND DREAMS

I was the first one in my family to make it to college so I know my family is very proud of me. Therefore, my family has a lot of hope in me, for me to succeed. My parents had struggled and suffered to leave Vietnam, for my siblings and I to come to America and get a good education, and I don't want to disappoint them. Especially my dad, he was a soldier in the Vietnam War, he was helping America fight for South Vietnam's freedom. After South Vietnam lost, my dad was put in internal [sic] camp for five years. It was very hard for everyone in the family and we were very poor. Therefore my siblings did not continue their education and had to work to help the family survive. After all the struggle and suffering my family went through, I feel that it is my responsibility to make their dreams and hope come true. It is also my dreams and hope that I'm working on to achieve.

—Cordelia, Asian American female immigrant student

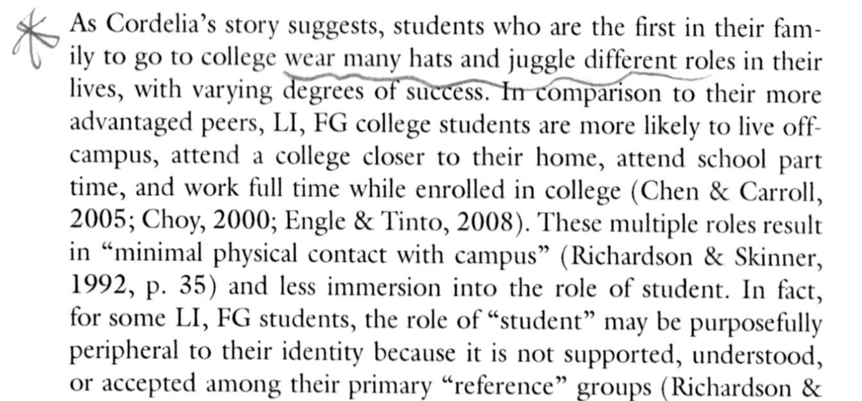

As Cordelia's story suggests, students who are the first in their family to go to college wear many hats and juggle different roles in their lives, with varying degrees of success. In comparison to their more advantaged peers, LI, FG college students are more likely to live off-campus, attend a college closer to their home, attend school part time, and work full time while enrolled in college (Chen & Carroll, 2005; Choy, 2000; Engle & Tinto, 2008). These multiple roles result in "minimal physical contact with campus" (Richardson & Skinner, 1992, p. 35) and less immersion into the role of student. In fact, for some LI, FG students, the role of "student" may be purposefully peripheral to their identity because it is not supported, understood, or accepted among their primary "reference" groups (Richardson & Skinner, 1992, p. 35).

For many LI, FG students, family obligations take center stage in their lives. These obligations play out differently with regard to diverse family structures, cultural contexts, and expectations. For student-parents, caring for their children and managing their schedules in tandem with the needs of their family shape the extent to which they are able to connect with and engage in their role as student. It is not surprising then, that students with children tend to have lower college completion rates (Adelman, 1999). For other students, parenting is a surrogate role, one they take on as an older sibling or cousin, where the day-to-day working of an entire family's schedule hinges on their availability to care for younger children, perform household tasks, and provide transportation between school, jobs, and home for family members.

Many FG students of color are from immigrant families and cultural backgrounds where the role of family is central to one's life, and "housework and child care" (Ginorio & Huston, 2001) are expected and required. In some cases, as with Latino families, these obligations "fall on girls more often than on boys" (Buriel, Pérez, De Ment, Chavez, & Moran, 1998; Ontai & Raffaelli, 2004; Sy & Romero, 2008). In addition to specific household tasks, childcare, and work, LI, FG students, particularly from immigrant families, are often called on to serve as "cultural brokers" and translators on behalf of their parents and family elders (Padron, 1992). Students might escort parents to medical appointments, assist with completing insurance paperwork, or serve as translators between their parents and other social, government, or housing agencies (Kiang, 1992). This results in a role reversal that may be disconcerting but necessary for both the students and their parents. On one hand, these students are asked to be in charge, but on the other, they must account for when or why they need to go to campus or have their own schedules. One student captures this dilemma in the following words: "At home we cooked and did chores, even in the third or fourth grade. We didn't know we could play sports, dance, act, etc. . . . In my generation I couldn't have my parents go to conferences without me being there, especially when I was smaller, because English was hard for them. This really interfered with my learning because I was *almost* mad at myself as if I weren't modernized enough or hadn't tried hard enough to make it in this world. But then, I also found so much pride in my family and background" (Poua, Hmong American female student).

Parents who have no experience with college may not understand the demands and schedules of collegiate life and may resent their interference with family obligations (Fuligni, Tseng, & Lam, 1999; Phinney, Ong, & Madden, 2000). These obligations can then become a "major source of stress and concern, especially in conjunction with academic pressures" (Constantine & Chen, 1997). FG immigrant students may feel an inordinate amount of pressure to succeed in college because, in many cases, they carry with them the aspirations of their family and community. To succeed in this country through the portal of education is a justification for all that their family and community have experienced in leaving their homeland for a better life in America. Yet the very people who proudly talk about their son or daughter as a "college kid" may do little to shift home life and responsibilities to make room for this new role as a student.

Students who are first in their family to attend college must then navigate through a complex web of roles, identities, and obstacles to get to college, and this process of finding their way takes on new meaning when they arrive on campus. While these students make many efforts and sacrifices to enroll in college, coming to college also gives rise to questions and internal negotiations as they try to make sense of their place, identity, and fit in this new world. FG students, particularly students of color, often question whether they can manage the academic demands of college and because of this, enter higher education filled with self-doubt (Réndon, 1992). They also wonder if they will be able to find social footholds to create a sense of community on campus and grapple with how to survive "intellectually and emotionally in an environment that devalued one's contributions because of one's race and cultural background" (Lara, 1992, p. 67). In addition, because FG students often become representatives and models of their communities, failure in school is not just their own to bear but also a reflection of their families, as well.

Finally, LI, FG students who are also from communities of color have come through the educational journey experiencing discrimination. The experience of perceived discrimination and perceived cultural differences, particularly in comparison to their White middle-class peers, impacts comfort level, connection to faculty and peers, and willingness to explore new opportunities on campus (Nora & Cabrera, 1996; Saldana, 1994; Schmader, Major, & Gramzow, 2001; Zea, Reisen, Beil, & Caplan, 1997). Explicit or suspected discrimination influences the self-confidence of students of color who are also grappling with how to bridge the divide between their home and school worlds and how to manage the financial demands of college.

GOING THE DISTANCE

FG, LI students must overcome harsh odds to get to college. Yet despite these challenges, these students continue to aspire to a better life and count on a college education as their conduit to a brighter future. The number of LI, FG students applying to college continues to grow (Carey, 2005; Engle & Tinto, 2008; Ishitani, 2003); yet the challenges that they face in getting to college continue to shape, impact, and impede their progress during college, which begs the question, How can college enhance the experience of FG, LI students? And more importantly, why should institutions invest in these students? The chapters that follow will draw on student voices in conjunction with theoretical arguments to make a case for why we must engage these students and what approaches we can offer for how we might best do it.

In a time when quality, accountability, and fiscal restraint are the centerpieces of college and legislative agendas, why should the academic experience or persistence of FG college students be of concern? There are several reasons: First, if education is indeed for the public good of the country, then the wide disparities in educational attainment, particularly with regard to race, class, and ethnicity, need to be seriously considered in order to prepare the future citizenry for stewardship of the nation. Despite some gains in postsecondary participation, Hispanics have one of the fastest rates of growth in the U.S. population but are the least educated in the country (Ruppert, 2003). Similarly, LI students are less qualified for college and are participating in higher education at significantly lower rates than their middle- and high-income counterparts (Ruppert, 2003). With continued growth in loan versus grant funding and with states' budget shortfalls, less and less money is being funneled into higher education. As a result, this change most directly impacts the students who are least likely to come to college in the first place (Ruppert, 2003). Even when FG students do make it to college, their persistence is significantly lower than that of their traditional counterparts. As educators and institutions, we look at all the factors that constrain LI, FG students and consider programmatic interventions, resources, and policies that could minimize these constraints. Although these are important things to consider, we may do well to look at our LI, FG students from a different frame, as well. We have in our classrooms and on our campus students who have demonstrated resilience, an ability to tolerate ambiguity in contexts of lived experience and identity, and a capacity to navigate multiple worlds. Their aptitude to successfully communicate in multilingual ways speaks not only to their often literal command of more than one language but the ways in which their life experience have prepared them to *speak* the languages of their many worlds. These skills are well-aligned with learning and developmental outcomes for all students. Our role, then, is to find ways to the bring these facets of LI, FG students into their awareness as strengths and to consider ways in which classrooms and campuses can draw on these strengths in the construction of knowledge and community.

There are economic, intellectual, and ideological costs that higher education must consider in creating a space and place for FG students. The economic implications are self-evident in that retention is cheaper than recruiting students, but it most also be noted that successful graduates from diverse communities can positively influence their states' economies and speak for their communities' needs (Carey, 2005). The ideological costs are, in some ways, far more weighty and are linked back to the work of pioneers such as John Dewey and more recently, Ernest Boyer (1990a), who ask us to consider and reconsider the role of higher education in expanding and supporting the creation of new knowledge and opportunity by developing open, inviting communities of learning for all students.

Chapter 2

Strangers without Codebooks

Isolation and Marginalization

When I first came to college, I didn't even know that what I was going to do at college, or what would it be like. I was so afraid of college because of all the comments I got about college. People would tell me that it was really hard, that I really needed to be ready for college, and without college, you couldn't do anything. After high school, I was really unsure about what I was going to do with my life. My peers were so serious about it that it made me really nervous because I didn't know what I was going to do; it seemed that they were all set and ready. College seemed so important; I had no clue what it was all about. I just thought in order to get a job, you had to go to college. So I just went with it.

—*Mai, Hmong American female student*

I remembered how awkward I felt upon entering the room where the course was to be taught because I came in with nothing, and I didn't know what to expect. There was something about the course that made me feel tense. Maybe it was the work, the effort, and the people that made me realize what I was in for. I was nervous about the decision I made by taking this course because I was a freshman—I was lost.

—*David, Asian American male student*

In their own voices, Mai and David, both first-generation (FG) college students at a large, predominantly White midwestern research university, describe their concerns about college. Mai captures the confusion and apprehension about her impending arrival, while David describes his first day of college. Mai confirms that the road to college itself has been unclear and expresses fear about how she will measure up in comparison to others in her peer group. Both express anxiety and concern about this new place, and more importantly, their place in it. It might be argued that it is not uncommon for any college student to feel "lost" as a freshman. However, unlike their counterparts who come from college-educated families, Mai and David are lost in a different way. They are not merely lost in the expanse of campus; rather, they have arrived without the "codebook." Both must now traverse an alien landscape while simultaneously figuring out the rules and expectations, both implicit and explicit, which shape every facet of the collegiate experience. In addition, it is unlikely that they will find many familiar markers that reinforce their lived experience, further solidifying the concern that they have, indeed, come to this new territory unprepared (Jehangir, 2008).

The last two decades have seen a major change in the demographics of higher education, with increasing numbers of women, students of color, and students from low-income (LI) backgrounds going to college, many of whom are the first in their families to do so. Despite such gains in access to postsecondary education, FG college students remain at a disadvantage in terms of maintaining enrollment and attaining degrees, particularly in comparison to their peers whose parents attended college (Horn & Nunez, 2000; Nunez & Cuccaro-Alamin, 1998; Warburton, Bugarin, & Nunez, 2001). FG students are, for example, more than twice as likely to drop out of college compared to students whose parents have college degrees (Chen & Carroll, 2005).

A critical problem for FG students is the significant isolation and marginalization they experience on college campuses. This experience results partly from their challenges in navigating the social milieu but is also reinforced by a curriculum and pedagogy that does little to reflect their worldviews. Isolation in this social milieu is typified by the fact that the nature of the journey to higher education is quite different for FG college students than for their traditional counterparts. FG students are more likely than not to be Hispanic or Black, to come from low-income families, to be older, and to be foreign born (Ishitani, 2003; Warburton et al., 2001). Thus while most students grapple with the expected transition to a new environment, FG students' transition and isolation are heightened by other social factors, ranging from economic and language struggles to geographical, racial, and cultural adjustments. These struggles present themselves in the form of anxieties, a sense of dislocation, and challenges in navigating a cultural landscape often alien to FG college students (Lara, 1992; Pascarella, Pierson, Wolniak, & Terenzini, 2004; Rendón, 1992; Terenzini et al., 1994).

Isolation for FG students deeply limits their engagement, involvement, and persistence in higher education (Nunez & Cuccaro-Alamin, 1998; Pascarella et al., 2004). Moreover, this effect is heightened when they find themselves marginalized in the curriculum. Populations historically excluded from higher education, such as minority, LI, and FG students, also often find themselves excluded from the college curriculum. Adrienne Rich (as cited in Takaki, 1993) captures this experience when she writes, "What happens when someone with the authority of a teacher describes our society and you are not in it?" (p. 16). Students who are first in their families to attend college bring with them histories and experiences that have the capacity to inform and enrich the learning experience, and in doing so, make them part of the academic community. Instead, being excluded, silenced, and rendered invisible in the curriculum only further marginalizes and isolates them on campus (Jehangir, 2008). This reinforces the idea that what happens during college can be as debilitating or supportive to the progress of FG students as what they experience prior to arriving on campus.

To understand the nature of isolation is to recognize one's sense of being cut off from the whole, to be disconnected from the way in which the world functions around oneself, or to feel apart from the language, conventions, rituals, and norms that drive the day-to-day activities. This isolation can manifest itself in different ways, and the way in which it plays out is shaped by factors that may be most salient to the identity and self-perception of a particular student and also how he or she is perceived by others on campus. Students who are the first in their families to go to college often arrive on campus as outsiders. London (1992) highlights the exclusion of FG students on campus by drawing on the work of Weber (1968), who argues that "a specific *style of life* is expected from all those who wish to belong to the circle" (p. 932, emphasis in the original). In higher education, particularly the predominantly White four-year institutions, this circle is still shaped by norms attributed to White, male, and middle-class students. Status group membership is shaped then by how "language, social conventions and rituals for all kinds, patterns of economic consumption, understandings regarding outsiders, and matters of taste in clothing, food, grooming and hairdo" (London, 1992, p. 7).

For FG students, their worldviews are markedly different with regard to preparation for college, family roles and obligations, as well as their familiarity with campus rituals, symbols, and implicit modi operandi. Magolda (2001) argues that the rituals in various educational contexts are deeply reflective of the political, social, and cultural expectations and norms of those institutions, and by examining campus rituals, one can determine "whose interests are being favored and whose are being ignored . . . who is at the cultural center and who is on the margins?" (p. 6). For FG students, many of whom are students of color, immigrants, student-parents and students with disabilities, the conventions of day-to-day life on campus render them to the margins, and the isolation of campus life is coupled with racial or ethnic isolation that they may not have anticipated.

Marginalization in the Curriculum and Beyond

The experience of isolation for underrepresented students is reinforced by the marginalization they experience in relation to college curricula. The challenges of being poor and a student of color are heightened by the fact that the cultural capital of higher education often does not acknowledge or value the cultural experience of these first-generation college students. Many students "felt like strangers to the classroom because the curriculum had nothing to do with their lives" (Rendón, 1996, p. 19). Still other Black students related that "American history was incomplete without the inclusion of African American experience" (Rendón, 1996, p. 19). London's (1996) research also reflects the need to develop continuity between academic culture and the culture of the students who participate in it. He notes the necessity of "making connections between educational practices and students' external lives and identities" (p. 13). For FG students, these experiences of isolation and marginalization collectively weave through classroom, co-curricular, and home life.

In particular, for FG students who are students of color, this sense of isolation and marginalization is played out in ways that reinforce their second-class status on campus. Ananda, an African American female student, describes encounters with overt racism in her residence hall: "I think when I first came here, freshman year, I went through some weird experiences. When I lived in the dorms, I had some people writing racial slurs all over my door and stuff like that. And I went to the university police station; they didn't do anything about it. And so I was really scared, and I didn't know what to do, and after that, I didn't know what to think about this university anymore."

Ananda's experience reinforces the many nuances of isolation that first-generation students of color face at predominantly White institutions. If they are already questioning their place on campus, then acts of overt or covert racism simply confirm and feed into doubts about their ability to be seen as legitimate participants in college.

THE COMPLEX INTERSECTION OF ISOLATION AND MARGINALIZATION

Welcome to College: Come Be Like Us

The ways in which isolation and marginalization impact students are deeply rooted in the structure of higher education. Students are welcomed to campus; they are invited to participate in a variety of programs that are often explained to them in words that make no sense, and acronyms that make even less sense. Not surprisingly, research has demonstrated that FG students are less likely to find a welcoming environment on campus (Terenzini, Springer, Yaeger, Pascarella, & Nora, 1996). For FG students, this invitation to participate often translates to an expectation to change themselves and adapt to the rules of academia. It is true that the process of adaptation to any new place, work site, or geographic setting may be useful and necessary for any person in order to make sense of a new world and learn to navigate within it. Yet for FG students, this process of adaptation requires a reinvention of the self, and oftentimes this demands cutting off, disguising, or undermining parts of their racial, ethnic, religious, or class identity. The extent to which this happens depends on the students' own development, the culture of the institution, and the discipline in which they might be working. Each institution in higher education has its own culture based on values, artifacts, and assumptions that, particularly at predominantly White institutions, are largely based on White, male, and middle-class norms. Institutional values are communicated implicitly and explicitly in the mission, goals, and communications of a given campus. Artifacts are the rituals, symbols, language, and stories that illustrate and operationalize the values of each institution. Assumptions are at the core of institutional culture because they underlie the day-to-day practices that reinforce these values and artifacts (Bensimon, Neumann, & Birnbaum, 1989; Kuh & Whitt, 1988; Richardson, 1994). Often the academy takes prides in having a diverse student body and espouses multicultural values, but it communicates nothing or little about what historically underserved students themselves might offer to the academy: "And herein lies the problem—while many institutions articulate values that espouse inclusivity and multiculturalism, the continued power differentials suggest that there is a disconnect between explicit values and operative assumptions" (Jehangir, 2009b, p. 294). The message that students receive is that their cultural capital, language, and resilience are not of use here; rather, they must reshape themselves in the likeness of the *status quo*, and only then can they be successful. In fact, as much as higher education problematizes the nature of inequities in public life and the role of higher education in addressing them, the structure and function of our campuses, disciplinary homes, classroom spaces, and programs often perpetuate the very things we wish to dismantle. Jarod, an FG White student, describes the role this inequity plays in limiting his educational experience:

> It is kind of like this, [higher education] is a whole completely different culture that talks a lot about the problems . . . and talks about these theories of the elite. I don't really know the right word. But it talks a lot about what's wrong, but yet they kind of still are that problem in a sense. I mean everything that we were studying in these courses (i.e., referring to sociology classes) that was kind of problematic, academia still represented. So it was kind of two sides. [The message] is we're going to teach you how to be a scholar. We're going to teach you that . . . who you are and who you were before, this isn't good. You have into change this.

All students will inevitably change throughout their formative years in college. Indeed, we hope that the educational experience does encourage growth and change, but for FG students the metamorphosis is often a result of forced necessity and an act of survival.

Words Matter: Do You Speak My Language?

For FG students, communicating in higher education is a process that requires learning a new language and giving up significant parts of themselves in order to make room for the "university way." What we in the academy may not consider are the ways in which the "centrality of language is a marker of difference (both in the university and at home)" (Law, 1995, pp. 5–6). To understand how to navigate the university, FG students are trying to do the same things that all students are trying to do: find their classrooms, connect with peers, adapt to new workloads, and interpret the expectations of different professors. The difference for them is that not only have they had less preparation and familial guidance to prepare them for this experience, but they must also grapple with the new language of the academy. For instance, How do I ask a question? How do I approach the financial aid officer? What happens the first time I am told "no" and get redirected to another person for an answer? What was that word she used? How should I respond to being questioned? Laurel Johnson Black, a working-class academic, recalls this internal struggle with language in her own undergraduate career. She says, "I remember clearly the first time I chose to say to a professor, 'Really!' instead of the more natural (to me), and what others might think of as more colorful, 'Gett outta heyah!'" (1995, p. 23). Black's comments underscore the experience of being in an environment where one sees little reflection or validation of their world, which makes the very process of giving voice to your ideas, questions, thoughts, and concerns daunting. Even when FG students do take that necessary risk of speaking, writing, questioning, or reacting, the prerequisite is to make sure that it is rendered into the language of the strange new land. Say what you have to say, but don't sound like yourself. Grace, an adult African American female who is a parent of two, talks about her self-doubt with respect to asking a question in class: "I sit in the back, and there's like are 160 kids, and [the professor] is like way up there, and if I raise my hand, my voice, will it echo? What if I sound stupid? I'll just email him. Yeah [Laughs] I'll just email; I won't ask in class, you know."

As such, language and voice are closely related in that one's capacity to cultivate voice and share ideas, particularly in new and uncertain contexts, is determined by the extent to which one feels equipped with the right "words" and if one expects to be correctly interpreted and heard by his or her audience. These issues of finding the right words to communicate, trying to be heard and making sense of academic language are compounded for FG students who are immigrants, refugees, and nonnative speakers of English. In many cases, their language skills may be proficient enough to allow them to serve as "interpreters and intermediaries between their parents and American society" (Kiang, 1992, p. 101). However, on campus the language barriers, terminology for which there is no translation in bilingual dictionaries, and concern about perceptions others might have about their accents create constraints, both within and outside the classroom (Harklau, Losey, & Siegal, 1999; Harklau & Siegal, 2009; Kiang, 1992; London, 1992; Rendón, 1992Roberge, Siegal, & Harlau, 2009). In both writing and speaking, students are thus in the process of trying to navigate their own voices through academic standards, varied audience expectations, and new texts in "a struggle for no less than their own identities" (Blanton, 1999, p. 136).

Britzman (1991) writes about the complex interplay between language, words, the speaker, and the audience:

> The struggle for voice begins when a person attempts to communicate meaning to someone else. Finding the words. Speaking for oneself and feeling heard by others, are all a part of this struggle. While tone, accent, and style qualify meaning, meaning is never realized by the

> individual alone. The struggle originates with the individual, is shaped through social interaction and mediated by language. Voice suggests relationship: the individual's relationship to the meaning of her or his lived experience, and hence to language, and the individual's relationship to the other, since understanding is social. It may be sparked by personal intent, but voice is always negotiated within context and situations, and by the meanings of others. The struggle for voice then, is always subjective, dynamic, interactive, and incomplete; it is never a matter of mechanical correspondence between the speaker's intentions, the language and the listener. (p. 23)

Britzman's insights about the ways in which language and voice are rooted in one's lived history and shaped by social interaction are central to the isolation and marginalization that FG students experience. Unearthing the subtext in communications with peers and faculty is also part of this challenge. When your story and narrative are not reflected, heard, or valued in the world around you, you are forced to attend to that reality. And this awareness brings caution and doubt concerning your place in this world.

Adjusting to Academic and Social Life

In addition to engaging in the process of translating the language and expectations of higher education, FG students must also learn how to balance the demands on their time, and they must prioritize their various roles as student, worker, and family member. Part of this journey involves deconstructing certain images and narratives that they might have come to believe about college life in order to come face-to-face with the realities and demands of being a student. Many students, including FG students, welcome the freedom from the confines of the tight schedules they were limited to in high school. Rashida, an African immigrant female student, writes about how her perception of college life was shaped by television shows and by talking with peers. She writes,

> My perception on college was a naive one because college was not what I thought it would be. I used to think that college was everything that I saw on TV—the partying and students slacking off in classes. I thought that everyone went to college so that they could have somewhere that they could party. When I was in high school I could not wait to get to college because I would have freedom to choose my classes and not worry about standardized tests that the government gives to high school students. I believed that my hard work in high school would finally pay off because I was now a college student and it is my turn to party and slack off.

For Rashida and other FG students, this idyllic sense of quickly gives way to the harsh realization that this hard-wo dom" is not free at all and requires a great deal of organizati and time management. Rashida speaks to this process of adjustment:

> When I got to college my perception was destroyed because college was not what [I] thought it would be. I did not know that there would be papers due every other day. I had to learn to organize and balance class with jobs and still be able to do my homework. I was in a world that worked in a fast pace and I had to catch up, if I wanted to pass and succeed in life. I realized that the students that were slacking off and going to parties were students that could *afford* to slack off and party. College is every expensive and I believe that, if you are paying for the class why are you wasting your money by slacking off and party all the time.

Rashida's words articulate both the challenge of balancing multiple roles and keeping up with the new rigors of college life. FG students come into to college less academically prepared than their second-generation peers. They are also less practiced at time management and study skills (Bui, 2002; Cabrera, LaNasa, & Burkum, 2001; Chen & Carroll, 2005; Lohfink & Paulson, 2005). Remedial or developmental coursework has been one response to students who are academically underprepared when they enter college, and 25 percent of students who attended a four-year institution between 1992 and 2000 completed at least one developmental course at the postsecondary level (Wirt, Choy, Rooney, Provasnik, Sen, Tobin, 2004). Despite these interventions, students who completed any remedial courses were less likely to earn a degree or certificate than students who had no remediation. In addition, when compared to their traditional counterparts, FG students reported fewer hours of studying per week, completing fewer credits in their first-year, and spending more hours per week working off-campus (Terenzini et al., 1996). To manage these gaps in educational preparation and to balance their multiple life roles, FG students are less likely to be engaged in academic and social experiences that foster success in college (Engle & Tinto, 2008). These extracurricular and co-curricular experiences, ranging from study groups and faculty interaction to clubs and organizations, are spaces that are critical to creating a sense of connection, belonging, and place for students, particularly on large campuses. FG students who are in most need of a "hook" that draws them into the academic and social life of campus are also the students who are least available to participate in these activities, given their competing life roles (Astin, 1997; Billson & Terry, 1982; Cabrera, Nora, & Castaneda, 1992; Nunez

& Cuccaro-Alamin, 1998; Pascarella et al., 2003, 2004; Terenzini et al., 1996; Terenzini, Cabrera, & Bernal, 2001). Just by hearing how Grace, an African American student, describes her weekly schedule, it is clear that her life roles demand all of her time, and the limitations on scheduling her classes could also impact timely progress toward her graduation:

> My first few years, I did 16 credits and I worked 3 jobs, while I had my daughter because she was in school so it made it easier. But now, I quit two of the jobs and kept one of them. Now I go to school Monday through Thursday and I work Friday through Sunday. I go to school till three everyday and then my daughter is in school till three everyday so it works out, and he [my partner] keeps the baby and he works at night so it's all set schedules like bam bam bam, so I don't have time to do other things with other people so, its school, work, school, work.

Grace describes her days as a ping pong game between work and school, with parenting and partnering thrown in. She has mastered a system that works for her and her family. That she will succeed seems likely; her determination to progress efficiently toward a degree speaks to commitment in the process. Nevertheless, her process will be different and may be lonely—she may not develop friends or study partners to call when her child is ill, and she may have to delay a class in order to keep her work schedule and finances intact. In fact, she may not graduate in four years. Her story is an example of many FG students who find their own ways to adjust to the academy, and when they are "delayed," it impacts their bottom line, as well as the rankings of the institutions they attend. More often than not, the institutional response has been to raise admission criteria to make it more challenging for students like Grace to attend a four-year institution. This structural gate keeping not only perpetuates the isolation of the few FG students who do make it in, but it also creates a world where the resilience and drive of students like Grace is falsely packaged as failure.

Curriculum, Classroom Dynamics, and Faculty Relationships

The classroom is where students spend the largest amount of focused and consistent time on campus. It is here where they have an opportunity to interact with peers and faculty, and how they perceive their place in the classroom impacts their perception of their status and value in higher education itself. The classroom dynamics are shaped together by the instructor, peers, the disciplinary stance, curriculum, and pedagogy.

Similar to language, this is a complicated dance, with performers at varying skill levels and cognizance of how the identities of all participants impact the whole. Richardson and Skinner's (1992) study of 107 interviews with FG students of color from 10 different public universities revealed incidents whereby low expectations of students of color held by faculty resulted in assumptions about their knowledge and capability to be successful students. This perceived bias on the part of faculty suggests that "preconceptions of the limited potential of minorities . . . can turn into self-fulfilling prophecies, restricting the achievement of all minorities regardless of preparation" (p. 34).

The process of learning and engaging with peers and faculty is also shaped by the extent to which students feel that their ideas, input, and lived experience matter. Bensimon (1994) points out that "while knowledge and theories are generated from the standpoint of particular interests, locations, and life experiences, we have been schooled to believe that knowledge is objective, neutral and separate from the knower" (p. 23). This framework of neutrality fails to acknowledge the power differentials that are at play in the classroom and on campus, not only when issues of race, class, gender, religion and nationality are part of the discourse but also in the insidious ways in which they shape the structural environment of higher education (Darder,1997; Hogue, Parker, & Miller, 1998; hooks, 1993).

When these students' lived experience is brought into the classroom space, by way of either discussion or circumstance, it requires them to sometimes unwillingly "out" themselves as representatives of their community or their group. Shani, an African immigrant female student, describes an experience of being asked or expected to speak on behalf of her religion: "I mean there have been moments where we'll be discussing something Muslim and suddenly everyone in class will look at me waiting for me to contribute or say something. There hasn't been any sort of genuine seeking of information at all . . . oh well."

She notes that this inclusion is perfunctory and additive at best and does little to validate her place in the classroom. This process of tokenism requires FG students to traverse the fine line between filtering their own feelings and thoughts with the realization that their voice is going to be heard and absorbed as a representative of their community, and as such, their words will have ripple effects on other students of color, immigrants, or nonmajority groups in the classroom and beyond. Davu, an African immigrant male student, captures this process of weighing the repercussions of sharing his authentic voice: "When I am sitting in Political Science class, and it's almost 150 people and one lecturer, and you're the only one, of two African Americans, you don't want to raise your hand. You feel like a an outcast in a sense; you raise your hand and say something that offends almost 95 percent of the audience. And I *see* myself, literally, I want to say something, I don't want it to come out in a negative or in a, in a bad way in front of people, or I don't want the Professor to look at me [negatively]."

As Davu's comments suggest, he is challenged not only by how his words will be perceived by a predominantly White peer group and faculty but also by which words he might use to effectively communicate his opinion. He goes on to talk about suppressing his ideas rather than trying to negotiate this difficult terrain of language and context, especially when he is aware that his words could shape the perceptions and attitudes of the class toward others students of color. He says,

> I could have expressed all my opinions, but it doesn't make a good image not only about myself when I am one of only three people of color. You have to always be thinking about how there's always a second or third aspect to how you say something. You always have to worry about how people, you know, how you preface it to most people: People, people of color, minority people, the people of the topic that you are responding to. You have to keep a good image in the classroom, because when you are like one of the three people of color, you always have to. Professors expect you to respond, but at the same time, I just look at the professors [and think], "You don't want me to say what I think." There are some topics where the professors come at you and be like, "So Davu, what do you think about that?" and I just sit there and be like, "I really don't know what to tell you." Then you don't express your deepest, fullest thoughts, but you tend to just give them your small ideas . . . it's always like that.

Not all interactions with faculty are negative, but given that many FG students feel like outsiders on campus and have heard the message that college faculty may not value their input, the willingness to trust and engage with authority figures is an issue. When interactions in class reinforce bias, prejudice, or racism, students are less likely to participate or trust that it will be different the next time. Thus how students are received in the context of the classroom dynamics, as well as what shapes the curriculum in class, can contribute to including or isolating FG students.

CAN'T GO HOME AGAIN

While transition to college is a time of change and adjustment for all students, for traditional students, these changes are part of a "normal rite of passage" (Jalomo & Rendón, 2004, p. 39), while for FG students, this journey requires redefining oneself in the context of this new world. Indeed, breaking away from parents and striking out is a central adventure of young adult development; in fact, much of the college experience aims to be a process that engages students to grow, change, and develop in ways that are new and different from their old high school selves. Yet for FG students, this separation process is both an "internal and external process that involves cultural dislocation and relocation" (Jalomo & Rendón, 2004, p. 39; Levy-Warren, 1988). Through interviews with FG students, London (1989) suggests that "family role assignments and separation dynamics [are] at the center of the drama of first-generation students" (p. 147).

This separation is some cases, is literal; it is a geographical move from home to campus. In some cases, it is a move to new cities and new states, which removes FG students from the familiar context of home and the cultural enclaves of their communities. This shift can be both exhilarating and terrifying at once, where students revel in their newfound autonomy but also come to realize the burden of managing new responsibilities and expectations without the proximity of family support or encouragement.

This separation is also rooted in negotiating one's place, both at the university and back home. As students engage in new experiences and grapple with new intellectual ideas about the world and how they see themselves in it, they begin reenvisioning themselves, as well. At an intellectual level, some students describe "a growing understanding of the workings and consequences of class and racial discrimination and how that understanding [changes] their lives" (London, 1996, p. 11). This process of intellectual investigation can lead to a sense of transforming their self-concept in a manner that requires "'leaving off' and 'taking on,' a shedding of one's social identity and the acquisition of another" (London, 1996, p. 12). The process of engaging with ideas and concepts that deconstruct one's past is not easy. As Olivia, a Native American student, notes, FG students who have also been challenged by racism and classism often feel the need to disguise their circumstances, and this "imposter syndrome" follows them well into their college years: "Racism and classism is something I understand, I have been in two [social] classes and basically two families over my 19-year-old life. One of the two classes that I was raised in was lower class that I was born into with my mom, and we had to be on food stamps and welfare. At Christmastime, my sisters and I would wrap up some of our old things just to make it look like we were getting a lot of presents when I had friends come over."

This process of cultivating a new sense of self brings students into internal conflict with their private and public identity. It is one thing to experience marginalization, but quite another to deconstruct it and make sense of it in one's daily experiences. Rita, an Asian American immigrant student, addresses this dilemma when she says, "Of course, as an immigrant I know how hard it is to achieve your dreams without losing yourself in it. Meaning, usually if you are able to compromise your identity with the popular culture of America, then usually you will get accepted quicker than if you think about your culture and values and identity. The old immigrants do expect the new immigrants to assimilate into their culture because I know of many people who don't like it when people follow their old culture."

As Rita's words suggest, FG students might feel the need to disguise their authentic voices or stories on campus in lieu of fitting in, but this problem may also extend to their life back home. Thus for FG students there is a sense of being an impostor in one world and a traitor to the other.

Many students feel a deep sense of loyalty to their family values, cultural norms, and expectations but also find that their new knowledge might place them at odds with these views. Clearly, this journey toward a new identity is a painful one and often leaves them at odds with their "home world" (London, 1996). One student described the feeling of loss created by this newfound knowledge and sense of self: "Sometimes family members get threatened or friends get threatened; they think that now you're in college, you think you are better than them" (Rendón, 1996, p. 18). Consequently, many FG students are challenged to get all that they can out of their college experience but try, sometimes unsuccessfully, to balance that out with family roles, expectations, and ways of life. As Lauren, an African American female student notes, the college world inevitably seeps into home life, and oftentimes the two do not mix well: "By this time of the semester, I was started coming in late and missing classes. Family and marital problems began to arise as I struggled to make it to class. I was living out some of the conflicts that were being addressed in the classroom, and ironically I was doing the best I could to take a stand for my individuality."

Other students recall how failures to meet parental, familial, or cultural expectations result in conflicts between their roles as students versus members of their families. The consequences of these conflicts can take a heavy toll on FG students. Anna, a Latina student, describes her feelings after a fight with her mother, who expected Anna to skip class and take care of an ill sibling so that she could go to work. Anna's reticence to take on this role resulted in both anger and disappointment on the part of her mother. She reflects on the challenges of being caught between her home life and her role as a student: "I kept thinking about myself and my life right now, and how I treated my mother the other day. We had a misunderstanding, and I screamed at her for the first time in my life. I felt like I was all alone and had no support. I felt really lonely, and I didn't think she understood."

This process of renegotiating relationships between family and friends, of "testing cultural ties and family codes of unity" (Jalomo & Rendón, 2004, p. 40; Rendón, 1992; Rodriguez, 1982) is agonizing for many FG students, who come to college with hopes of making their communities proud and creating lives that are better and more secure than the ones their parents lived. Yet often the clash of their two worlds results in an existence that leaves them on the fringe of both the academy and their home world. Rendón (1992, 1996) argues that while moving toward healthy individuation is important, slow movement allowing students to maintain ties to family and culture is vital for many FG students.

FG students often walk a tightrope between their home and school worlds, feeling apart from both. While even the most well-meaning families of these students find themselves ill equipped to help them make a successful transition to college, we in higher education can no longer claim ignorance. Osei-Kofi, Richards, and Smith (2004) argue that "we have the opportunity to begin creating and nurturing environments where multiple forms of knowledge, identities, locations, and ways of knowing hold credence" (p. 56). This process of identifying new ways to think, teach, learn, engage, and be engaged by all students requires reflection on who we think belongs in the academy. If we still hold to the ideal of education as a public good, then we must reexamine how access, excellence, and inclusion can be mutually dependent. The next chapter explores critical pedagogy and multicultural education as theoretical frameworks from which to consider these questions in the context of the FG student experience.

PART II

BEING HERE

SURVIVING THE TRANSITION TO COLLEGE

Chapter 3

Reimagining the University

Theoretical Approaches to Serving First-Generation College Students

As already addressed in the previous chapters, first-generation (FG) students are not a monolithic group; indeed, their multiple identities and life roles have both constrained and enriched their perceptions of educational institutions and their place in it. In the United States, higher education is still viewed as the great equalizer, the path to a better life and a means of providing a more stable future for one's family. In spite of this hope and promise, the question of who should have access to higher education has been debated in different forums for centuries. In 1965, John Brubacher, a higher education historian, raised the question "Higher education for whom?" (Hurtado & Navia, 1997, p. 106). While, in theory, one goal of higher education has been to redress imbalances in society by creating access to college, the practice of this goal has been in question, largely because access and excellence have been viewed as mutually exclusive. We have still failed to reach real consensus on this issue, even in 2010.

The question of entitlement versus access to higher education is particularly relevant to public four-year institutions, charged with serving the citizens of their state while simultaneously at the mercy of state funding as decided by the legislature (Jehangir, 2002b). Faced with the pressure and seemingly contradictory goals of excellence and access, public institutions have struggled to articulate and carry out the promise of real access at four-year universities, where resistance comes not only from the outside but also from within the ranks of academe.

U.S. higher education is unique in its diversity. It is an amalgamation of two-year, four-year, public, private, tribal, and community colleges and universities, each with different missions and goals. In spite of these differences, most would agree that across institutional type and size, there is a broader commonality: the mission of serving the public good and preparing students for a sense of citizenship and participation in their communities. Despite these commonalities, there are broad philosophical differences regarding the practices of teaching, creating community, and educating young citizens in higher education.

Two issues are at the core of this debate. The first issue raises the question of who belongs in the academy, particularly in four-year institutions. Connected to the tussle between access and excellence is the second issue: disagreement on the content and relevance of the college curriculum today and the role of higher education in serving the needs of the changing student body. This changing collegiate body is typified by the browning of America and includes not only more students of color but also more women, more students who are multilingual, large numbers of part-time students, adult learners, immigrants, and FG college students (Carey, 2005; Rendón, 1994; Takaki, 1993). It is not difficult to see that related issues are tied together such that there is a push-and-pull dynamic between efforts toward inclusivity, both in content and student body, and efforts to maintain a sense of academic freedom and specialization in disciplinary fields.

In this chapter, with attention to the context delineated above, I draw on a body of scholarship from multicultural education and critical pedagogy as the theoretical frameworks that explore and deconstruct the ways in which underserved students are isolated and marginalized in higher education. Given my arguments about the isolation and marginalization of FG students, the discussion and intersection of these two theoretical frames serve as the basis for the perspective I use in discussing the obstacles that FG student experience and the undervalued potential of the contributions they bring to higher education.

The intent of this chapter is not to delve into the complex nuanced and rhetorical differences between and within the two, but rather to locate points of synergy and to draw on the ways in which they complement each other. Finally, I will also highlight why these frameworks in praxis are salient to challenging the isolation and marginalization of FG students on campus. This chapter begins by situating these theories in the context of higher education and addressing the ways in which the structure and function of our institutions are often contrary to creating an engaged space and place, not only for FG students, but for many who work and study within the institutional system.

Tierney (1997) argues that "public education has a responsibility greater than admitting those who score highest on a standardized test. Public higher education is a public good" (p. 192). But to whom is this good bestowed? Today's admission and financial aid processes are impacted by the fiscal constraints of the institution, as well as its strategic efforts to build an incoming class of students that enhances rankings and status (Bowen, Chingos, & McPherson, 2009). Even when underserved students matriculate at these institutions, scholars have argued that admission to college is only a small part of the access dilemma (Alderman, 1999; Engstrom & Tinto, 2008). While more FG students might be admitted to college, the structural, curricular and ideological ethos are slow to evolve in ways that effectively engage students who have historically been underserved.

Certainly, within institutions there is evidence of pockets of innovation. In some cases, broader administrative commitment in the form of both ideas and dollars has resulted in the creation of curricula, programming, interdisciplinary border crossing, and learning partnerships that aim to create inclusive, engaging spaces for all students, including underserved students. The intent of this chapter is not to dispute these efforts, but to note that they are often what Betsy Barefoot (2000) calls "innovations on the fringe"—that is, they impact those who participate but are limited in challenging the status quo. The theoretical frames discussed below aim to challenge the status quo and to address how effectively institutions "guarantee conditions necessary for all equally to further their aims and by how efficiently they advance shared ends that will similarly benefit everyone" (Levin, 2007, p. 47).

As college campuses grapple with becoming democratic multicultural communities, there is no one definitive road map on how to get there. There has been significant discussion, however, about how validating curriculum, student and faculty connections, and peer relationships can impact students' social and academic engagement and invite institutional belonging to historically marginalized students (Hornak & Ortiz, 2004; Rendón, 1996, 1992; Tinto, 1997).

Boyer argues that "campuses should be viewed by both students and professors not as isolated islands, but as staging ground for action" (Boyer, 1996, p. 148). In this vein, both critical pedagogy and multicultural education are theoretical frameworks, philosophical stances, and methodological approaches that draw attention to inequities in education and reenvision campus and classroom spaces to be informed by the lived experiences of students (Gay, 1995; Giroux & Simon, 1989), particularly those who have been historically underserved. The next sections will briefly address the philosophical underpinnings of critical pedagogy, multicultural education, and the complementary connection between the two theories, concluding with their particular relevance to FG college students.

CRITICAL PEDAGOGY: DEFINITIONS, STANCE, AND CONTEXT IN TEACHING AND LEARNING

Critical pedagogy raises questions about "the influence of race, class and gender (and their intersections), how power relations advance the interest of one group while oppressing those of other groups, and the nature of truth and the construction of knowledge" (Merriam, 2002, p. 10). Hegemony is central to critical pedagogy, where power, sometimes blatant and other times masked as neutrality, shapes and directs the standards and content of literacy while limiting the experience of historically underserved students in educational settings. Grounded in the liberation struggles of Latin America, critical pedagogy is associated with the works of Paulo Freire and Che Guevara, to name a few. It has seen growth in the United States due to the efforts of scholars such as Henry Giroux and Michael Apple (Sleeter & McLaren, 1995).

Key concepts in critical pedagogy include resistance, oppression, transformation, power, domination, liberation, voice, and empowerment (Gay, 1995). These are frames through which critical pedagogues express critiques of policy, institutional climate, and curriculum in schools by reinforcing that "any discussion of public schooling has to address the political, economic, and social realities that construct the contexts that shape the institution of schooling" (Giroux, 1992, p. 162). This philosophical stance is aimed at consistently linking inequities in education with social problems in society at large (Giroux & McLaren, 1989). As such, it brings attention to the ways in which educational institutions fail to embody the ideals of democratic egalitarianism and social and economic mobility that they commonly espouse.

In the context of teaching and learning, critical pedagogues see their role as facilitators who encourage *meaning making* in their instruction, which is contextualized in the sociopolitical realm of the students. Specifically, this pedagogy "seeks to draw out student voices and put these voices into dialogue with others in a never ending cycle of meaning making characterized by reflection/action/reflection/new action and so forth" (Rivera & Poplin, 1995, p. 223). This process is an intentional effort to challenge current educational practices and curricula that give "unfair advantage to the cultures, traditions, and experiences of the European middle class" (Gay, 1995, p. 163).

Bringing the voices, stories, and the lived context of historically marginalized students (such as FG students) into the classroom "challenges teachers and students to empower themselves for social change, to advance democracy and equality as they advance their literacy and knowledge" (Shor, 1993, p. 25). Acknowledging and valuing the functional and cultural literacy that students bring to higher education confronts what McLaren (1998) calls the "pedagogical unsaid" (p. 45); that is, the ways in which both pedagogy and curriculum can perpetuate the dominant culture and it's values and ways of being normative. The effort to critique the educational system and draw parallels between inequities in the classroom and those in the larger society is a means of giving voice to silenced issues and creating avenues for students and teachers to be empowered agents of change. The praxis of critical pedagogy is rooted in naming intricate relationships between racism, classism, and power and challenging how these relationships create webs of constraint for historically marginalized students and their families within the educational system and beyond. From the perspective of teacher and student, critical pedagogy at its best invites both parties to collectively deconstruct these "isms" and engage in reflexive active participation toward changing these systemic inequities.

MULTICULTURAL EDUCATION: EVOLUTION, IDEALS, ENACTMENT

Multicultural education was born out of the civil rights movement in the 1960s, when the push for a more inclusive curriculum resulted in the first steps to creating courses and materials that reflected African American and other ethnic communities in the United States (Banks, 2001b). Since the 1960s, many different approaches have been created, critiqued, and redefined in an effort to address the changing student body and educational inequities in the United States. Since its conception, multicultural education has been through a variety of stages that reflect the slowly changing ideology and paradigm shifts related to difference and equity in education in the United States. Currently, the term *multicultural* is used widely with varying intent and meaning: conservatives and liberals exploit the term to serve their agendas, and despite different stances in both groups, "difference is tolerated—even celebrated as long as it does not contest" predominantly White normative values (Sleeter & McLaren, 1995, p. 14). The following brief discussion of the various approaches to multicultural education highlights the ideological shifts in its praxis and concludes with an approach that guides the theoretical framework of my work.

Early expressions of multicultural education, referred to as "teaching the exception and culturally different," sought to assimilate nontraditional students into the "standard body of knowledge and set of values and skills that all American students need to acquire"

(Sleeter & Grant, 1988, p. 43). Critiqued for perpetuating a culturally deficient worldview, this approach was followed by a "human relations approach" that relied on antiprejudice training and stereotype reduction to promote "unity and tolerance" (Sleeter & Grant, 1988, p. 75). While efforts to challenge stereotypes are of value, this interpersonal approach minimizes social inequity to the level of the "inability of people to get along" rather than really challenging inequities in the status quo; as such, it lacks the teeth to effect real curriculum change (Sleeter & Grant, 1988, p. 99). The human relations approach is often used in conjunction with other approaches, but by itself, it does not effectively articulate the connection among issues of diversity, class, and power differentials.

A third approach to multicultural education is what Sleeter and Grant (1988) refer to as "single-group studies," where the focus of elective course content is on specific ethnic groups, and students enrolled in these courses are typically students of that ethnic group. The initial focus on elective courses with specific ethnic group content was related to the women's movement, thus creating a voice and place for women's issues in the collegiate curriculum. Riding on this momentum and supported by social concerns for human rights in the 1970s, other ethnic groups, people with disabilities, and the Gay, Lesbian, Bisexual, Transgender (GLBT) community also gained some presence in the curriculum (Banks, 2001b; Sleeter & Grant, 1988). The fourth approach is referred to as the "multicultural education approach," whose advocates argue that the true ideology of multicultural education is "one of social change—not simply integrating those who have been left out into society, but changing the very fabric of that society" (Sleeter & Grant, 1988, p. 139). Critics of this approach argue that, in addressing cultural and ethnic differences, this approach does not effectively address social class or disability issues, nor does it pay enough attention to structural inequities or the tools that students will need to challenge them.

The last of the five approaches, referred to as the *multicultural and social reconstructionist approach*, best represents the tenets and theoretical framework espoused in this book. The term *reconstructionist* calls for a reexamination of the effectiveness of major institutions in meeting the needs of today's society and suggests that past systems "developed during the preceding centuries of the modern era are now incapable of" addressing the oppressions and societal inequities between different groups of people today (Bradfield, as cited in Sleeter & Grant, 1988, p. 175). The reconstructionist philosophy applied to a multicultural educational setting seeks to transform "the whole educational process," rather than merely creating new programs or pedagogy within the confines of the current system (Sleeter & Grant, 1988, p. 175).

This approach, also sometimes referred to as *emancipatory pedagogy*, *critical pedagogy*, or *transformational education*, argues for the necessity of teaching "political literacy" (Sleeter & Grant, 1988, p. 177). The practice of this final approach to multicultural education involves four dimensions. The first is practicing democracy, which means that education should help students practice democracy by creating opportunities for them to develop their own ideas, challenge authority, work together, and debate one another. Banks (1981) articulates this sentiment when he writes, "Opportunities for social action in which students have experience in obtaining and exercising power should be emphasized within a curriculum that is designed to help liberate certain ethnic groups" (p. 149). The process of empowering students is reflected in the second dimension: analyzing the circumstances of one's own life. This principle argues for the importance of students' self-examination and analysis of their own experiences so that they can develop constructive responses to societal injustices, rather than engaging in destructive forms of opposition. The third and fourth dimensions involve developing social action skills and coalescing. The former suggests that the educational experience is a training ground to prepare students for "socially active citizenry" (Sleeter & Grant, 1988, p. 191), while the latter articulates the necessity to understand and develop coalitions across race, class, gender, and other group divisions.

For the purpose of this book, I distinguish between diversity and multiculturalism in the following way: Diversity is the recognition of the "existence of different social group identities," including race, class, gender, sexual orientation, age, home language, disability, and religion (Miksch, Bruch, Higbee, Jehangir, & Lundell, 2003, p.6). Diversity is not interchangeable with multiculturalism: "If diversity is an empirical condition—the existence of multiple group identities in society—multiculturalism names a particular posture toward this reality" (Miksch et al., 2003, p. 6). Banks (2001b) suggests that out of all these different schools of thought, multicultural education is at least three things: "an idea, an educational reform movement, and a process" (p. 3). This conceptualization is captured by Grant and Tate (2001) who define multicultural education as "a philosophic concept built upon the philosophic ideals of freedom, justice, equality, equity, and human dignity that are contained in important documents such as the Constitution and the Declaration of Independence. It recognizes however, that equality and equity are not the same thing: equal access does not necessarily guarantee fairness" (p. 145).

With attention to multicultural and social reconstructionist education, this chapter argues for the necessity and value in linking this approach with critical pedagogy to move away from compensatory add-ons in curriculum and campus life, toward a more transformative, emancipatory vision of education (Ogbu, 2001).

CRITICAL PEDAGOGY THEORY AND MULTICULTURAL EDUCATION: COMPLEMENTARY FRAMEWORKS

Critical pedagogy theory is a strong parallel to multicultural education theory and is particularly germane to unpacking the experience of FG students because it provides a theoretical basis for the educational contexts of historically marginalized students. It looks at educational settings and curricula not as value-neutral, but rather as "symbolic property—cultural capital—which schools preserve and distribute" (Apple, 1990, p. ix). Given that the premise of creating the Multicultural Learning Community (MLC) with the dimensions of identity, community, and agency was specifically designed for students who have existed on the fringe of academic life, it is important to include a theoretical framework that questions the limits of "legitimate knowledge," as approved by the dominant culture.

Clearly there are many definitions of multicultural education (Banks, 1992; Gay 1995) and critical pedagogy (Apple, 1990; Giroux, 1981). The intent of this section is not to debate the subtle nuances of differences in these schools of thought, but rather to demonstrate the commonalities within these definitions and to illustrate the ways in which critical pedagogy and multicultural education theory form important linkages in the praxis of teaching and learning. It has been argued that both critical pedagogy and multicultural education theory are mutually supportive and share a common goal of action that is addressed by contesting historically established perceptions of knowledge in both theory and practice (Gay, 1995; Rivera & Poplin, 1995). Sleeter and McLaren (1995) articulate the work of these two frameworks in this way: "We do not consider either critical pedagogy or multicultural education to consist simply of a set of methodological formulations. Rather, both refer to a particular ethno-political attitude or ideological stance that one constructs in order to confront and engage the world critically and challenge power relations" (p. 7).

Thus neither critical pedagogy nor multicultural educators view their work as simply additive to the current curriculum. Instead it is an active process of challenging the perception of "legitimate knowledge" and reductionist pedagogy, which "emphasizes not the construction of meaning but the acquisition of meanings constructed by others" (Rivera & Poplin, 1995, p. 224). By contrast, the intent of critical and multicultural pedagogies is to engage participants in a process of knowledge construction, within the context of understanding the sociopolitical issues of their environment and their own experiences.

Kincheloe and McLaren (2000) argue that "critical research can be best understood in the context of empowerment of individuals. Inquiry that aspires to the name *critical* must be connected to an attempt to confront injustice of a particular society or public sphere within the society. Research thus becomes a transformative endeavor unembarrassed by the label *political* and unafraid to consummate a relationship with emancipatory consciousness" (p. 291). Critical pedagogy is closely connected with multicultural education theory and is a worthy lens in this study because it recognizes that, contrary to traditional positions, educational institutions may "actually work against the interests of those students who are most needy in society" (Darder, 1995, p. 329). In recognizing that educational experiences are deeply impacted by economics and access to cultural capital, this framework acknowledges that the isolation and marginalization of FG college students shape their learning experiences. Recognizing the historic contexts and hegemonic forces that limit educational success for needy students, critical pedagogy also argues that multicultural curricula can be created to reflect the "cognitive, motivational and relational styles" of nondominant students (Darder, 1995, p. 334). Also, the classroom experience can serve as a learning community to integrate activities and opportunities that honor students' stories in ways that allow them to gain voice and empowerment.

THEORIES IN PRAXIS: CURRICULAR TRANSFORMATION AND BRIDGE BUILDING

Jane Tompkins (1990) has argued that the "classroom is a microcosm of the world; it is the chance we have to practice whatever ideals we may cherish" (p. 656). Often these ideals reflect a meritocratic approach to teaching and learning. Supporting or validating "weak" students or engaging students' life experience is viewed as hand holding or irrelevant to the "academic" goals of the class. Rendón (1994) challenges these notions of education and suggests that "validation is an enabling, confirming and supportive process initiated by in- and out of class agents that foster academic and interpersonal development" (p. 44). Building on this, I suggest that when students feel validated in any learning context, they are more likely to engage in the classroom with a degree of vulnerability, risk, and effort, resulting in deeper and more authentic learning for them and their peers. Rather than "a separatist view of teaching and learning . . . that detaches the student from what is being learned" (Rendón, 2009, p. 35), both multicultural education and critical pedagogy argue for reshaping classroom spaces so that students, particularly those who have been historically marginalized, are invited into the learning experience as co-learners and co-teachers. By recognizing the cultural capital that these students bring to the table and by incorporating their voices in context with course materials, we create a laboratory that reflects the world beyond the classroom. Together, teachers and students can engage in knowledge construction employing multiple ways of knowing—ways that acknowledge that who people are as individuals shapes how they know. There is no one way to engage students and their lived experience into the classroom; likewise, there is no one formula that works every time. Despite this challenge, I draw on some key concepts from critical pedagogy and multicultural education that are particularly salient to engaging FG students. Student narratives are included in each section to demonstrate how their learning experiences illustrate the value of these conceptual approaches.

LIVED EXPERIENCE

> *Working in linked courses helps relate the materials to real life. The more links you have in between courses, the more links you start to see elsewhere in your everyday life "courses." Isn't that one of the purposes of higher learning: something you can use in life, not just in the classroom? It helps my learning because it opens my mind to links I should be making in the world.*
>
> —*Rose, African American female student*

Curricular and co-curricular spaces that value the lived experience of their participants is a critical component to create validating spaces for FG students. Validation pays attention to what Parker Palmer (1998) calls the inner and outer landscape of teaching and learning. The inner landscape gives attention to "subjectivity, intuition, emotion and personal experience," while the outer landscape is "associated with intellectualism, rationality, and objectivity" (Rendón, 2009, p. 7). I use the term *lived experience* to acknowledge that one's life encompasses past experiences that are interpreted in the context of implicit and explicit messages from individuals' environments. People's educational narratives are thus shaped by how they make sense of their experiences in schooling within the context of family, peers, culture, race, class, and gender expectations (Van Maanen, 1990). Two students from the same high school do not have the same experience any more than two students who get the same grade in a course have the same set of skills or weaknesses. Attention to lived experience contextualizes the learning experiences of students because it draws on their expertise and situates their ways of looking at and interpreting the world in the context of and as part of the knowledge construction process. This process asks to what extent teachers prepare students to understand, analyze, and explore how "implicit cultural assumptions, frames of reference, perspectives and biases influence the ways in which knowledge is constructed within it" (Banks, 2001b, p. 4). The argument can be made that instructors, regardless of discipline, could engage their students in this process. For many FG students, the acknowledgement of their ideas, experiences or beliefs within the learning process has been absent or muted. Tina, a White FG student, captured the implicit assumptions and biases within her high school experiences when she said, "I was never asked what I thought, and it was never important that everyone hear what I had to say."

Curriculum and pedagogy that allow students to bring their lived experience to bear on their learning capitalizes on their cultural capital and validates them as knowers and teachers. The use of lived experience can be reinforced by curriculum and assignments allowing students to connect who they are with what they are studying. While there is an assumption that this type of learning is best suited to the social sciences and humanities, faculty in science, technology, engineering, and mathematics (STEM) have also been successful in inviting students to bring their narratives into the classroom (Duranczyk & Lee, 2007; Solomon, 2009; Staats, 2005).

The use of lived experience, narratives, or stories as a part of learning can be perceived as "fluff." That is, for some, it is a nice addition, but to view it as critical to the learning process dilutes time and attention away from "real" learning. I have had discussions with colleagues who feel that academic discourse is not to be confused with students' telling stories because the latter only creates a therapeutic environment that undermines the intellectual project at hand. I argue that much of this has to do with how one facilitates and invites student ownership of the classroom space. Creating opportunities for students and teachers to share narratives from lived experience is a good precursor to building a sense of trust and belonging in the classroom. Doing so can break down barriers of race, class, ethnicity, nationality and other differences and can allow students to move from superficial discussions of topics to a place of deeper engagement (Chang, 2002; Jehangir, 2010a). Brad, an African American FG student, expressed the value of hearing the stories of his peers in their individual contexts: "I felt most engaged in class this week when we were doing the self-identity discussion. I really enjoyed being able to hear about everyone's culture and their stories. I felt that everyone had a significant amount of information to share with everyone else. No two people are alike, no matter if you're both Black, White, or yellow. I like when we do things like this because it gives me a better opportunity to get to know a little bit of everyone in class."

In addition, creating spaces for students to express their ideas and draw from their lived experience to substantiate these ideas allows their expertise to be valued. It also pushes other students to see one another as knowers. Sasha, an African American FG student, described the value of learning from her peers and noted that their multiple identities inform her in different ways:

> I feel that the first five weeks were a test to see if college was the place for me, and it is. I felt in those weeks that I was not alone. I find that I was able to learn and grow with my classmates. Over the course of those weeks, I realized I learned as much from the students as I did from the instructors. Sometimes hearing things from them was easier for me to understand than from the teachers. I learned from them, we all come from different places with different backgrounds, and that has an impact on what we feel about issues that were presented in class.

Sasha's comments speak to the way that sharing stories, beliefs, and perspectives in the context of academic topics and materials can push students to digest information in ways that aid in meaning making. For FG students, this process is a means of seeing themselves and others as contributors to knowledge construction rather than as merely bystanders.

Ananda, an African American FG student, described how the inclusion of lived experience in the curriculum and pedagogy impacted how and what she remembered and reflected on as a writer and thinker: "Instead of just learning about this stuff, we got to hear students' different experiences behind it. We got to hear family stories about migrating here to America, or about people moving from different regions or just . . . everyday stories. I think stories stick with me more than just a lecture. So I think that helped. A lot. It made it easier especially for papers because we were able to incorporate those experiences students had in there, and then tie it in to sociological topics and stuff like that. So, that helped a lot."

Ananda described how incorporating narratives in learning moved her from defining conceptual ideas to applying them in the context of real-world issues (Jarmon, Brunson, & Lampl, 2007). The diverse stories of her peers provided a lens through which she could make sense of specific sociological frames, but they also reinforced the notion that these conceptual ideas have relevance to the lives of students such as her own.

Curricular materials that intersect with the lived experience of diverse students are another way of reinforcing their place in the academy. Anna, a Latina FG student, described her connection with a reading assignment that deconstructed myths and stereotypes about Latina women. Her assignment created an opportunity for her to understand the aspects of her own lived experience that she had previously felt, but could not name. She wrote,

> Anyway, I never truly cared about how I saw myself racially. But this article has helped me to identify with that part of me. As the author talked about the stereotypes of being a sexy and fiery Latina, I just nodded my head in amazement about these very ideas about the portrayal of Latina women, and how they are so real in my life. I just thought they were personal struggles that only women in general can relate to. However, the dress she described and the attitudes behind them have ironically been a part of my past. I enjoyed being able to label the experiences I've had; sexual advances, expectations, and role limitations. Though I know that all women experience these things in some form, the unique Latin myths and stereotypes have been a part of my life all along, and now I can actually claim that part of me. Wow, I didn't think that would ever happen in a class.

Anna's revelation demonstrates that being able to see one's self in the curriculum and finding overlap between the world within the classroom and one's identity challenges the isolation that many FG students might feel in the classroom. Multicultural educators argue that *content integration* in the curriculum (Banks, 2001a, 2001b) involves a conscious examination of the extent to which educators, classroom space, and policies reflect a variety of cultures and groups. It addresses the extent to which instructors use many cultural perspectives to illustrate key concepts, principles, generalizations, and theories in their subject area.

MULTIPLE IDENTITIES

Both multicultural education and critical pedagogy acknowledge and name the multiplicity of what students bring to the educational enterprise. Levin (2007) notes that the role of "student" is but one component of the identity of nontraditional students. Similar to traditional students, their identities have other components such as parent, child, Muslim, Hmong, girlfriend, and employee, to name a few. For all students, these differences are "negotiated, co-created, situated and socially constructed, involving not only one's understanding of one's self but also a social discourse about oneself" (Levin, 2007, p. 66). The challenge for FG students is that these identities and the demands of each may often be in conflict with each other, and students may find that being in the new world of college exacerbates these conflicts, with little opportunity to reflect or make sense of them. The capacity of teachers and peers to acknowledge this multiplicity is not only validating for students; it can also be normalizing if students' see that others are wrestling with similar issues. In addition, acknowledging the multiple identities of students can allow them to reframe their seemingly disparate selves as strengths. Students can begin to see their capacity to cross borders between worlds and selves as a skill and not a weakness.

Shani, an African immigrant student, addressed how recognizing and being able to name her many identities was a critical moment in her learning during her first year of college. As a senior, she reflected back on the value of contextualizing herself within the academic content of the class:

> It was interesting because so much of the discussions that we had . . . we gave meaning to the readings by talking about ourselves. I have never had an academic space where I could talk about the Hijab, or my identities a Muslim Somali woman, or even my identity in this country as a refugee. So, I guess it's something that I just have never had in an academic setting. I mean, certainly not high school. I'm glad that I was able to experience it so early on in my college experience, and it's just something that I am not shy to talk about.

Other students have actively practiced masking their identities, both literally and figuratively, to fit in. Jennifer, an Asian American FG student, described her struggles to fit in. Her experiences within a class with a multicultural curriculum allowed her to move from external formulas of perceived success to reflecting on and honoring her multiple identities. She wrote,

> I learned how to accept myself and express my mind and just took a step back and started to show my real self. Accepting myself, I also accepted my heritage and where I have come from. I was able to understand that I am a young Asian woman in America, and that I should not

> have to conform to wanting so much to be someone I am not. I used to wear blue contacts, never wanting to date someone who was Asian and even trying to dye my hair really light brown. From taking this class, I was able to see all these things that I have been doing and incorporating them with my life and how I can change.

Both Shani and Jennifer spoke to how curriculum and pedagogy can attend to their multiple selves. Shani referenced specific opportunities to speak to her own identity, while Jennifer's reflections pertained to how the collective impact of the curriculum and lived experience of her peers allowed her to examine and deconstruct her own identities. As such, Ross (1993) suggests that the role of a multicultural critical pedagogue is not to attempt to represent every cultural group in the curriculum, but rather to "construct identity allegorically in order that each group is able to see his or her cultural narrative in a broader and more comparative relationship to others and within the larger narrative of social transformation" (Sleeter & McLaren, 1995, p. 17).

CONFLICT AS A CATALYST TO LEARNING: PROBLEM POSING AND THE HIDDEN CURRICULUM

Engaging with the lived experience of students is messy territory. Instructors do not get to choose which stories or beliefs will be shared, and how they will be received. They do, however, have the opportunity to cultivate a classroom environment that allows for this ambiguity to be a useful tool in teaching and learning, as it is in life outside the classroom. To acknowledge the multiple identities of students is to open a door for them to engage as "whole" learners; however, doing so also creates conflict. While academic discourse has prized debate and dissension to further knowledge, this practice requires participants to engage in discourse with intellectual rationality and personal distance. Disagreements are often a zero-sum game with winner and losers.

By contrast, classrooms that seek to embody the tenets of multicultural education and critical pedagogy draw attention to social inequities and power differentials by relating conceptual ideas to lived experience in order to help students practice democracy. When students are pressed to develop their own ideas, challenge authority, work together, and disagree with one another, education becomes an "opportunity for social action in which students have experience in obtaining and exercising power . . . within a curriculum" (Banks 1981, p. 149).

Aisha, an African American student, addressed the value of disagreement in the classroom and how it empowered her and moved her to examine dissension in a broader context:

> This week I really got engaged in the discussion over the case of Separate but Equal. I think that it was great that I got to tell everyone how I felt about it. This discussion was tense because everyone was yelling something out and going against each other. The discussion really made me think about the fact that the world has changed because if it wasn't for the challenges to [the] Separate but Equal Act, I wouldn't be sitting in the same class with all the kids in our class. I think the discussion is what really got me hooked.

It is not surprising that this process of connecting learners to content personalizes it in ways that both engage and create disequilibrium. The disequilibrium that students' experience can take many forms: internal dissonance, discomfort concerned with identity issues, and tension that builds when learning to negotiate working with others. Curricula that actively engage diverse FG students with issues of racism, classism, sexism, homophobia, ethnocentrism, and other practices of exclusion begin to name and deconstruct issues that have directly impacted their lives, their schooling, and potentially, their futures. Understandably, this new knowledge can be disconcerting and difficult to absorb. Jennifer, an Asian American student, commented on these challenges:

> During the first two weeks it was difficult for me to understand and accept everything. I was not allowing myself to be educated in every article that we were reading because I did not want to learn all the truth about the subjects. Once I learned everything, it was too much for me. The truth about what every race has gone through to come to this country was too harsh for my understanding because throughout my life, I have had this façade created in my mind about all of the subjects. I learned and come to realize that I was a television child. I was raised by both my parents, but however, I learned most of my things from television.

Jennifer's comment captures that initial sense of being overwhelmed in the face of a new realization of social inequities. This engagement with one's discomfort is central to what critical pedagogues refer to as *problem posing*. Problem posing aims to "bring interactive participation and critical inquiry into the existing curriculum and expand it to reflect the curriculum of the students' lives" (Wink, 2000, p. 60). The process of asking questions that one would rather not address and to posit responses to these questions is to move beyond the visible curriculum and begin to uncover the ***hidden curriculum***. The hidden curriculum involves "the unexpressed perpetuation of dominant culture through institutional processes" (McLaren, 1998, p. 45). It is embedded in how historic issues and conceptual "truths" are framed and also in the language used to explain them. This perpetuation of normative values, images, and aspirations in schools and beyond is what Jennifer referred to when she called herself a "television child."

The hidden curriculum is as much about what is *not* said as it is about what *is*. When students in a college classroom express disbelief about U.S. citizens of Japanese descent being interned in camps during World War II or the immigrant experience in America, they are demonstrating how the hidden curriculum has shaped their worldviews. Shelley, an African immigrant student, addressed the internal conflict that arose when problem posing revealed "unsaid" (McLaren 1998, p. 45) facts about other cultural groups and immigrants in the United States:

> During week six, we were discussing our cultural backgrounds. The main thing that stuck with me was when Lian was talking about the Hmong immigrants and their struggle of migrating to America because of bad conditions they experienced in their homeland. She talked about how the Hmong avoided death by fleeing to the jungles. I was so shocked when she was talking about it because when I was in high school, I had friends that said the Hmong were taking all the jobs and they are over populating Minnesota. I did not know anything about Hmong people and their history. I assumed that they were in the same position as many immigrants, which I myself am. Immigrants moved to better their lives and become accepted in a society that believes in equal opportunity. The Hmong came to America expecting equal opportunity and acceptance; however, what they got was a harsh realization of assimilation. Since I was so caught up in making myself seem more American and fit in with my "White" friends, I did not want to accept my past. So when Lian said that the Hmong helped America fight in a war, I was so shocked that I did not know what to believe. This is a great learning experience that I will never forget, and I thank her for sharing an emotional story with us. Being in a class like this has helped me understand people better and realize that it is not all right to not know about other cultures, but to generalize based on people's opinions is not fair to that culture.

Despite internal conflicts that many students may experience as a result of delving into the lived experience and courses with multicultural content, this problem posing stimulates questioning and reflection that allows students to arrive at new understandings. Shani, an African immigrant student, discussed her reticence to consider inequities in the gay and lesbian community. Her previous beliefs were rooted in the dominant culture and religion, which dichotomized social justice and morality: "Long before this topic came up, I was honestly a bit nervous about discussing it since I'm a person of faith. But I've learned that my faith and my ideas on human rights can and do coincide with each other. This week has been wonderful for me, and I have learned a lot. I found myself questioning myself a few times, and I learned a lot about myself."

Other students described how conflict went beyond internal struggles and shaped their interactions with peers. Yet the process of engaging in conflict about ideas, beliefs, and positions also encouraged students to understand how to negotiate with one another and how to respect opposing points-of-view. Cooper, a biracial student, spoke to this process in his comments below:

> What do I remember the most? You know, I remember lots of arguments and discussions all the time. Which was a good thing, cause it gets you thinking about subjects you normally would never be concerned with. But there was lots of arguing. I think a lot of it also had to do with just being in the same three classes with the same people all semester long, so you really get to know everybody and feel comfortable kind of standing your ground, but in an educational way. I guess discussion is the better word, but . . . you know, there's sort of anger, and confusion. And you kind of like realize that things aren't as equal as you really think they are, and everybody doesn't think the same; we all have different opinions and views.

Conflict, or the act of engaging in disagreement, is part of taking ownership of one's learning and involves pushing students to process their ideas, hear alternate views, and give voice to their beliefs. Conflict can be a catalyst for learning when it allows for and acknowledges a bridge between feeling and learning, between anger and understanding, between individual and community, and between process and product. Conflict is not new to FG students; in fact, many have experienced conflict both internally and overtly as they have tried to bridge their worlds and identities. They have not, however, had many opportunities to name their conflicts, to be validated or challenged by the lived experiences of others, nor have they had many chances to process their disequilibrium.

SELF-REFLECTION AND MEANING MA

If conflict is critical to moving students forward in their unc
of themselves and the world around them, opportunities to
how they learn, think, and feel and create meaning is a centra piece of the puzzle. Laura Rendón (2009), a self-described FG academic "scholarship girl," describes her own schooling experience where she "rarely had assignments that asked me to reflect on the meaning and purpose of what I was learning" (p. 3). Her experience reveals learning as a separatist process where students "bank" (Freire, 1971) knowledge from instructors who are the sole experts. By comparison, critical multicultural learning spaces seek to cultivate spaces of "connected knowing" (Belenky, Clinchy, Goldberger, & Tarule, 1986), where learning fuses content with participation, relationships between teachers and students, and understanding oneself. As detailed in the previous section, this process of fusing interdisciplinary approaches and lived experience allows students to consider themselves anew in cognitive, interpersonal, and intrapersonal ways. If conflict can be a catalyst to learning, then opportunities to reflect on intrapersonal, cognitive, or interpersonal conflict are central to making meaning of these experiences.

Self-reflection with regard to how one thinks, feels, and engages with others in the college context is critical for FG students when bridging the divides between their multiple worlds and beginning to develop a new conception of the self. Student development scholars have referred to this journey toward a new personhood as *self-authorship* (Baxter Magolda, 2001; Kegan, 1994). Pizzolato (2003) defines self-authorship as "a relatively enduring way of understanding and orientating oneself to provocative situations in a way that (a) recognizes the contextual nature of knowledge and (b) balances this understanding with the development of one's own internally defined goals and sense of self" (p. 798).

As the definition suggests, when students encounter critical moments or disequilibrium, they are pushed to make meaning of this and ideally translate that meaning into an internal sense of self that helps them navigate new information, relationships, and problems. Many FG students have experience dealing with provocative situations where they must bridge the chasm between home and school life or encounter institutionalized racism and classism in other milieus. By contrast, they may not have had many opportunities to reflect, digest, interpret, and address how they cope with these experiences. If the curriculum and pedagogy deconstruct students' lived experience in the context of academic content, there must be space for students to absorb and grapple with the dissonance they experience to progress toward self-authorship. Assignments, classroom exercises, and writing that cultivate self-reflection can capitalize on disequilibrium and can provide a safe outlet for students to express ideas and feelings for themselves.

Kelly, a Native American student, talked about the role that reflective writing and journaling played in deepening her understanding of herself as a student: "Reflective writing really enhanced my learning. In a way that I have something on paper and can visually remember and work on my strengths and weaknesses. As I reflect, I am more aware of myself and how I learn. I can restructure an environment when I learn from what adds to my learning and takes away at the same time."

Her comments seemed to answer the question, how do I know? and spoke to a deeper cognitive understanding of her learning style preferences.

Shani, an African immigrant student, addressed how she engaged in the process of meaning making and recognized that her identity was not a closed book. Rather, her intrapersonal development was being shaped by the intersection of her previous beliefs, values, new interactions, and knowledge in college. She was working to answer the question, Who am I?

> Throughout this semester, I learned that I don't exactly completely know myself—and this is OK. I'm constantly changing my views, I'm learning new things, and I'm trying to make sense of everything that's going on around me. I surprised myself a lot this semester. I always felt that I had a perfect image of what kind of person I was, and now I realize that I can do this only to an extent, but I've also got to open up a little and explore my horizons. I can't always have set rules in my head and expect everything to follow through according to certain personal beliefs I have, or precedents.

Sasha, an African American student, reflected on the intrapersonal dimension of self-authorship: "As I make my mistakes, I'm slowly progressing through them. The errors that have occurred so far in my life have made me a more wise and knowledgeable person. Some mistakes happen more than once because sometimes certain mistakes take more time to cure. It's like certain scars take [a] longer time to heal."

Each of these students reflected on their learning and their lived experiences in different ways to address and make meaning of the different facets of their identities. While the outcome of this reflection may, in fact, raise more questions, it is the *process* of becoming consciously attuned to one's thinking, feeling, and engaging with the world that is especially important for FG students. This attention to and valuing how one makes meaning of learning allows students to better understand, navigate, and contribute to their college experience. Diane, a Hmong American student, talked about how early experiences with reflection became part of her toolkit—something she could draw upon to deepen and make sense of much of her college career:

> The reflection part of this whole program . . . like how we were, not forced, but that was one of our assignments to reflect on readings or things discussed in class. I find myself doing that, even though I don't have the need to do it for any kind of class. I still find myself journaling, blogging . . . things like that in my life. I think that when I do that, I capture confusion, and things that don't really make sense to me, and then when it does, and I read back to it, it's like, "I learned that lesson or this lesson about myself." So it's encouraged me to reflect and just learn from experiences and really take time to jot down the thought process that I go through. Which is really important to me because that's what helps me define who I am.

As these students' voices suggest, reflection can be employed as a "contemplative practice tool" (Rendón, 2009, p. 69) that can serve students to better understand themselves well after a class is over.

Shared Learning and Intellectual Reciprocity

Classrooms can and often do perpetuate hegemonic practices of the dominant society. These "institutionalized social relations of power" (Darder, 1991, p. 35) are expressed in the curricula, pedagogical approaches, student-teacher interactions, peer relationships, and policies and procedures of the school itself. Yet the classroom is also a bounded space that allows instructors the opportunity to engage students in dismantling this hegemony. Creating spaces where students are invited and then expected to teach and learn together is one way to challenge the "banking" theory of knowledge. When students become co-instructors, they can reach one another and take fellow students to places that an instructor may be unable to. Both Nadeah and Olivia discussed how hearing stories of their diverse peers helped them relate to and normalized some of their experiences with racism. Nadeah, an African American student, found that she learned from other students when they expressed their racial and cultural backgrounds in their own words: "I could probably relate to everybody's misfortunes, and I liked how everyone had a story to tell about their childhood experiences with racism. And I really enjoyed getting to know some of the backgrounds of the students when we talked about our races and heritages."

Olivia, a Native American student, connected with another biracial student and understood how she felt displaced in her own identity. She wrote,

> Something that I also related to was when Rachel felt that she did not fit in with White or Black kids in the class. And I also felt the same way when I was a child when I used to go to pow-wows because I was too White compared to other the Native American children. I did not really fit in with a lot of the White kids because they thought I was White, but when they found out that I was Native American, it made them look down on me. Where I am from, people think that Native Americans are trashy, drunks, drug users, and poor.

While sharing one's stories and ideas may not be easy or painless, for students to hear one another was a step toward recognizing how their identity or life experience had been framed by the dominant culture. Rosa, a Latina student, described how hearing from her peers, absorbing their perspectives, and weighing how they made sense of the world added to her knowledge fund (Moll, Amanti, Neff, & Gonzalez, 1992). She wrote,

> Well, I think that I learn something new every day, so when I come to school and we discuss our readings, and when we talk about questions as a group, I like to think that I learn a little bit about everyone each time we do this. I like talking about questions that were assigned to my group because I feel as if I learn a lot from their way of thinking when we discuss these questions. A way I also found of learning is that I like to write down things when we are discussing the questions in class. This way I not only have what I thought was the answer, but I have a little bit of what everyone thinks.

Her comments suggest that she was engaging in learning that recognizes that there are many ways of knowing and that the "truth" may be relative to context. Bao, a Hmong American student, extended on Rosa's comments by acknowledging how valuing diverse and contradictory perspectives on given issues can enrich how one comes to understand issues more deeply: "I thought that the discussion on Thursday was very good, that students [brought] out good points that some may not agree [with], while others agree. My classmates are really up to what they have to say. I think that sharing what we think makes other people understand how we feel and by doing that, people can learn and teach each other."

Grace, an African American student, commented on how shared learning with her peers opened her eyes to the points of connection and struggle in their collective stories. For her, this intellectual and affective engagement shifted how she viewed and valued her peers and moved her to a collective view of struggle and agency:

> It was easy to act like you want to bloom, but you must remember that it takes time and a lot of attention to fully come to life. When I came here, I was self-indulged. By that, I mean I didn't think much about the feelings of other races and their struggles. Instead, I was blinded by my own biased opinions of other races. By the fifth week, my rosebud started to open. The in-depth conversations we have had as a class only stimulated my mind to learn more. I feel more of a closeness with other races whose ancestors have traveled almost the same journeys as mine, fighting for the same rights and privileges of a free man.

FG students have often had complex or ambiguous experiences particularly in context of educational settings. Many have experienced racism and classism, in both overt and covert forms, and have had to learn to straddle the border between worlds, languages, social contexts and varying expectations. This type of knowledge does not fall to any one discipline or area of study; rather, it is "integrative learning" based on real life experiences that draw on "multiple modes of inquiry, offering multiple solutions and benefiting from multiple perspectives" (Association of American Colleges and Universities and the Carnegie Foundation for the Advancement of Teaching, 2004), p. 1). When FG students are encouraged to draw on their funds of knowledge and to demonstrate how their cultural capital informs their peers and their academic projects, they move from the margins to a more integrated center in the academy.

VOICE

"To find voice is to discover a capacity to engage in self-expression and, in doing so, to construct knowledge. It is to engage in the world of ideas, concepts, and feelings, both cognitive and effective, and to find ways to articulate one's place in that world. Finding voice is irrevocably tied to the notion of self within community, because one's voice does not exist in a vacuum; rather, it is embedded in and impacted by context, language, and position of the speaker and by the community in which he or she must speak" (Jehangir, 2009a, p. 40). Many FG students describe how unpracticed they are at voicing their ideas, feelings, or thoughts in educational settings. For many, prior learning experiences have undermined or silenced them so much that when they experience curricula, instructors, and peers who value their input, they are surprised. While there is no one prescription that can speak to all students and encourage them to risk voicing their thoughts, classrooms that use lived experience and draw upon students' multiple identities to create safe spaces for reflection, constructive conflict, and shared learning can build the trust necessary for FG students to articulate their voices.

Jarod, a White student, described his surprise when he found this sense of trust and risked sharing his ideas:

> I spoke my mind. I had things I wanted to say, and to my utter shock, I actually said them. As I wrote in my learning log, "I actually had things to say." This was an important point in the class and possibly my life, in the sense that before that, unless extremely angry, I never felt comfortable speaking to others for fear of ridicule. The first five weeks of class was about me finding my voice, finding ways to let go of these uncomfortable feelings and becoming an important part of every single discussion. I said in more than one learning log, "I got my point across well." That became very important to me. I took it upon myself to always convey how I felt. In the first few weeks, this came as a big shock. Throughout high school, I always was the kid in the back pretending to listen. In conversations I had with family members, I can't describe their surprise that I was constantly talking in class and becoming an important voice in all my classes. My hard work, in turn, gave me confidence in my intellectual abilities. From reading the first five learning logs, I was trying to accomplish two things—establishing confidence in myself and find my "voice" in class.

Jarod's newfound confidence was derived from being in a validating space with peers who took interest in his contributions and demonstrated how he was cultivating a new academic identity—one where he was deeply invested and had ownership in his learning.

Diane, a Hmong American student, commented on how her sense of voice was shaped by a multicultural curriculum that encouraged her to speak from her position as a Hmong woman. The opportunity to step out of the dominant discourse and lay claim to her own identity while answering questions from her peers was empowering:

> The exercise on self-identity was a really good project, I thought. I've never been able to share with people my race and ethnic background before. It was cool to see people ask questions because the reason is they don't hear about Hmong people. Usually when people look at me they think I'm just Asian. The most dominant group would be Chinese. Then when I tell people I'm Hmong, the majority would say they have never heard of Hmong people. I thought this discussion was very engaging. As a learner, the discussion on race, ethnicity, and culture raised a lot of questions. Mostly questions toward myself. I remember asking myself why I didn't share this or that.

Diane also noted that this process of finding voice was never complete; rather, it gave rise to more self-directed questions and reflection. Xue, another Hmong American student, addressed how finding voice is not limited to verbal dialogue; it also includes ways in which people share themselves in alternate media. She described the process of sharing of herself, her narrative, and making herself vulnerable in the context of an art project: "At first I was so nervous about how my self-portrait was going to be, compared to everyone else's. But when I got to the show mine and put it up, I felt like I just released a whole new part of myself. I felt amazed with myself. I was so scared at first about my portrait, but then I realized, "Why should I be afraid: this is me, I am myself." After my presentation of my portrait, I felt the great, tense feeling rush through my body. The flow was relieving, I was so happy with myself."

When Xue spoke to that sense or relief she felt from sharing herself with others, she captured what it means to give voice to one's concept of self. Many students of color, immigrant students, and low-income students expend so much energy fitting in or cultivating a persona that they feel is appropriate for higher education that they cannot engage their authentic selves in the classroom. Taking the opportunity to cultivate and exercise one's authentic voice is an act of constant negotiation among language, words, the speaker, and one's audience.

Xue commented on speaking about this struggle to find her voice and how voice shifts in relation to context. Her comments speak to how FG students engage in give-and-take with their multiple identities and how they must find ways to speak with attention and awareness to the salient aspects of their identities in given contexts. Embedded in this process is the ability to negotiate not only with oneself but also with peers, instructors, and other stakeholders at the university. Voice, then, is about self-awareness and relationships with the self and others.

Both multicultural and critical pedagogy argue that in order to validate and learn from the voices of FG students, it is necessary to begin by acknowledging normative power with attention to how and to whom it is given in educational contexts. Attention to lived experience, multiple identities, and the role of conflict in making meaning are ways of engaging historically marginalized students to challenge these power structures and claim their space in higher education. It is also a means of making education more relevant to the experience of *all* students. Self-awareness and shared learning are both individual and collective processes that allow students to become attuned to themselves and to one another. None of these efforts to engage FG students are linear, formulaic solutions to strengthening their place in college, but they are an amalgam of factors that seek to address the many ways in which these students are isolated in higher-educational contexts.

When students see themselves in the classroom by way of curriculum, lived experience, and identity, they have opportunities to contribute their cultural wealth to the academy. Yosso (2005) refers to this wealth as "the total extent of an individual's accumulated assets and resources" (p. 78). FG students bring their aspirational, linguistic, cultural, social, navigational, and resistant capital to the academy. As crossers of social borders, FG students have fostered skills that challenge inequalities in employment and schooling contexts. They have navigated multiple roles, languages, and social institutions. Yet the academy does little to tap into this cultural wealth, and in fact, undermines its value in learning.

Critical multicultural pedagogy argues that engaging with this cultural wealth in the classroom puts students in the position of dismantling normative forms of knowledge and cultivating their own voice and place as legitimate participants and contributors to the academy. Some might interpret this focus on situating learning in students' lived experiences as indulgent and more concerned with social connections and camaraderie than real learning. Yet if we agree that knowledge is complex, multidisciplinary, and dynamic, then inclusion of the self and understanding the perspectives of others is critical to constructing knowledge. These approaches are not limited to FG students; all students benefit from experience dissonance, which prompts a deeper understanding of the self in the context of the world. However, FG students are bringing to the academy funds of knowledge long misunderstood or underutilized, so attention to these pedagogies is a means of bridging gaps that must be addressed in creating inclusive and representative communities in higher education.

While these pedagogies can be practiced in any classroom, the next chapter considers learning communities as a powerful vehicle to enact critical and multicultural pedagogy in undergraduate education.

CHAPTER 4

TOWARD COMMUNITY, CONNECTEDNESS, AND CARE

A good design cannot be piecemeal; multiple patterns of the design of a room must connect to the house, the garden, the neighborhood, the town. This is a fundamental view of the world. It says that when you build a thing you cannot merely build that thing in isolation, but must also repair the work around it, and within it, so that the larger world at that one place becomes more coherent, and more whole.

—*Christopher Alexander,* A Pattern Language: Towns, Buildings, and Construction

Alexander's quote (1977) speaks to the need for attention to connectivity and integration in designing space. This notion of connectedness suggests that no space is a vacuum; rather, it is an extension of the spaces that lead up to it, and that space embodies the experiences of those within it. The last chapter addressed the theoretical frameworks of critical pedagogy and multicultural education in cultivating a sense of place for first-generation (FG) students. This chapter considers redesigning the classroom and draws on learning-community structures and core practices as a means of operationalizing the tenets of these theoretical frameworks. Learning-community design is explored as a means of cultivating a space and place for FG students. The terms *space* and *place* refer to how students' locations or places in the world (Bruch, Jehangir, Lundell, Higbee, & Miskch, 2005) are impacted by their history, demographics, primary language, and their relative proximity to power. In this chapter, I share some history about learning communities and highlight key structures that impact learning-community design. I address research findings that focus on the impact of learning communities on the college experiences of FG students and end with a discussion of learning-community design as a vehicle for the praxis of critical pedagogy and multicultural education.

American higher education has been an amazing experiment of education as democracy. Yet despite the expectation that a college education is the first step to a new and better life, approaches to teaching and learning have made very small shifts in adjusting to the new generation of learners entering academia. In spite of increased access to college for FG, low-income (LI) students and students of color, college graduation continues to elude these groups of students disproportionately when compared to their traditional counterparts (Adelman, 2007; Renner, 2003; Roueche & Roueche, 1999). These concerns have raised questions about the value of current educational practices in preparing all students to engage in a global, multilingual, and technologically driven world.

In response to these questions, several reports have suggested the need for educational reform and curricular transformation that reengages both students and teachers into a collective project of problem solving, collaboration, and interdisciplinary approaches to create "deep" (p.2) learning (Kinzie & Kuh, 2004) that connects students with the world around them and with one another (Association of American Colleges and Universities National Panel, 2002; Boyer, 1996; Kellogg Commission on the Future of State and Land-Grant Universities, 2001). As the architect Christopher Alexander (1977) suggests, these efforts to create inclusive, transformative learning spaces "cannot be built in isolation, but must also repair the work around it, and within it, so that the larger world at that one place becomes more coherent, and more whole" (xiii).

As institutions, particularly large public universities, grapple with the new majority of FG students on their campuses, they are faced with the challenge of how to reshape themselves into sustaining democratic multicultural communities. While no one method, approach, or philosophy provides a definitive answer, there has been significant discussion about how "choices of curriculum structure and pedagogy invariably shape both learning and persistence on campus" (Tinto, 1997, p. 626), especially in ways that seek to "validate" (Rendón, 1994, p. 39) students' presence on campus and invite them into the broader learning community. The benefits of an integrated, participatory learning environment have been extolled by many higher education scholars, and classroom learning communities are one means of bringing together disciplinary, programmatic, curricular, and pedagogical approaches to connect disparate campus silos that do little to contribute to a cohesive learning experience for students.

Learning in Community

Seeing the classroom always as a communal place enhances the likelihood of collective effort in creating and sustaining a learning community.

—*bell hooks,* Teaching to Transgress: Education as the Practice of Freedom

As an instructor, I have struggled with translating the ideals of critical and multicultural pedagogy into my practice. This struggle is not uniquely mine, nor should it be surprising that it is difficult to do. In the previous chapter, I addressed ways in which attending to lived experience, multiple identities, conflict, and reflection are means of challenging the isolation and marginalization that many FG students feel in college.

Yet I know that these practices are not formulaic, nor can they exist in a void. They can only come to fruition in a space that actively invests in building trust, partnerships, and connections among all participants in the classroom. Learning communities are "intentionally designed classroom spaces that restructure curriculum and time and space for students" (Smith, 1991 p. 43), and in doing so, they can serve as vehicles for engaging in critical multicultural pedagogy. Asking any student, particularly FG students, to give of themselves, to construct knowledge, and to take the risks inherent in sharing one's narrative in the context of a classroom cannot be authentic unless there is trust and a sense of community. When I speak of "community," I do not mean the type of space where everyone agrees or even always likes one another. A community is a space where students and teachers practice a sense of caring toward the members belonging to the community; individuals work to take ownership not only of the class but also of one another. This type of caring offers safety and validation but also demands accountability and commitment. Janet, an African American student, captures this sense of community:

> The theme I had discovered in these first few weeks of class has been [a] sense of home. I have really made a bond with a bunch of people I had no intention [of] bonding with. I know I have begun a relationship with this group of people that would help enhance my talents and abilities and allow me to be myself. It is because of everyone's attributes and ability to work with each other; that's why we have that sense of home. I always refer to us as a family. Why? Because to me we act like family, we all have our differences and things that annoy each other. We all get on each other's nerves, some seem as if they don't do their share of work, and others

> seem like they do too much. Sometimes we want to complain about each other; we fight more often than not, but we can never stay mad at each other. And we always respect each other and we care about each other's feelings. I know, because it sounds a lot like my own family, that's how I can relate to some of the things that happen in our classes.

Cultivating this sense of community is intentional, effortful, and imperfect, but it is critical to the practice of transformative pedagogy. To this end, the learning-community movement offers pedagogical structures that work to intentionally redesign and integrate curriculum and pedagogy to create spaces to practice the learning community. The idea of learning in a community extends as much to teachers as it does to students.

Key elements of the learning-community design are akin to the praxis of critical multicultural pedagogy and can serve as a vehicle to combat the isolation of FG college students. Core practices in learning communities include: community, diversity, integration, active learning, and reflection/assessment (Smith, MacGregor, Matthews, & Gabelnick, 2004). Each of the core practices reflects values that resonate with the praxis of critical and multicultural pedagogy.

The concept of learning communities can be philosophically linked to John Dewey's principle "that education is most successful as a social process and is deeply rooted in our understanding of community and democracy" (Lenning & Ebbers, 1999, p. ix). Dewey articulated the benefits of student-centered, experiential education. Other early pioneers of the learning-community model include Alexander Meiklejohn and Joseph Tussman. In the 1920s, Meiklejohn created the Experimental College at the University of Wisconsin, in which students were required to take a set of lower-division courses over a two-year period (Goodsell-Love, 1999). Meiklejohn's living-learning community encouraged the importance of developing relationships between students and teachers. He argued that connecting curricular and co-curricular experiences and embedding present-day issues in the classroom enriched student learning. Closed in 1932 as a result of falling enrollment and dissension at the university about the nontraditional curriculum, its legacy is seen in the Bradley and Chadbourne learning-community programs today (Altschuler & Kramnick, 1999; Smith et al., 2004).

Tussman used a similar model in the 1960s when he created the Experiment at Berkeley, a two-year program in which student cohorts took a set of courses that had been preselected and taught by a team of faculty. This format of integrated learning was, in Tussman's opinion, a way to prevent fragmenting the student learning experience into separate classes of "horizontal self-contained units that were generally unrelated and competitive" (Tussman, 1969, p. 6). The philosophical ideals of access, democracy, and community are at the center of the learning-community movement; indeed, these core beliefs are closely connected with the goals of equity, advocacy, and curricular transformation in critical and multicultural pedagogy.

Cross (1998) argues that learning communities encapsulate and utilize "a changing philosophy of knowledge" (p. 4). Unlike the traditional view of knowledge, which suggests that the learner discovers external realities, the nonfoundational view of knowledge is built on the fundamental assumption of constructivism in which "knowledge is actively built by learners as they shape and build mental frameworks to make sense of their environments" (Cross, 1998, p. 5). This process of collaborative interdependent learning with other students and teachers suggests that there is a "growing sense that teaching and learning don't really happen unless there is some kind of building of relationships—not only between teacher and student but between teachers, students, and subject" (Palmer, as quoted in Claxton, 1991, p. 23).

Cross and Palmers' perspectives articulate a paradigm shift in higher education, suggesting that knowledge is socially constructed rather than an external entity that must be grasped. Rather, we need to create "environments and experiences that bring students to discover and construct knowledge for themselves, to make students members of communities of learners that make discoveries and solve problems" (Barr & Tagg, 1995, p. 15). To this end, Barbara Leigh Smith (1991) suggests that "learning communities work because they go beyond the individual and are a basis for coming together to build the 'larger educational enterprise.' They provide a powerful associative structure—a structure for bringing people together—in an environment with few effective centripetal forces" (p. 47). For FG students and for institutions themselves, learning communities can serve as sites of curricular transformation and can provide spaces to "develop pedagogical practices for diverse learners" (Lardner, 2003, p. 1).

LEARNING-COMMUNITY STRUCTURES

A focus on integrated, noncompetitive learning has been a core component of the learning-community models as they have evolved from the early 1960s to the present. The current learning-community literature is deeply connected with collaborative and cooperative learning theory, as reflected in the work of learning theorists such as Kurt Lewin, Martin Deutsch, and L. S. Vygotsky, along with their successors,

Kenneth Bruffee and David and Roger Johnson (Goodsell-Love, 1999; Lenning & Ebbers, 1999; Smith, 1991). Given the breadth and depth of the learning-community movement, it is not surprising that learning communities are defined in multiple ways. With a wide spectrum of institutional types and needs, learning communities can be flexible and can adapt to many different settings, student groups, and programmatic goals. A working definition of learning communities that captures their most salient characteristics states that they are "a variety of curricular approaches that intentionally link of cluster two or more courses, often around an interdisciplinary theme or problem, and enroll a common cohort of students. They represent an intentional restructuring of students' time, credit, and learning experience to build community, enhance learning, and foster connections among students, faculty, and disciplines. At their best learning communities practice pedagogies of active engagement and reflection." (Smith et al., 2004, p. 20)

Clearly, the literature streams on active collaborative learning (Bruffee, 1999) and cooperative learning (Johnson & Johnson, 1991) are closely tied to the curricular models of learning communities. In fact, the characteristics of many learning-community models incorporate collaborative and cooperative learning techniques and group expectations. While there are different learning-community models in higher education, all are typically characterized by small learning environments for faculty and students, building integration into the curriculum, employing active learning approaches, acclimatizing students to college expectations, and supporting students in establishing academic and co-curricular networks in college (Shapiro & Levine, 1999; Smith et al., 2004).

While these characteristics typify most learning communities, there are many different types of learning-community models, and scholars have organized them into multiple formats (Goodsell-Love & Tokuno, 1999; Lenning & Ebbers, 1999; Smith, 1991; Smith et al., 2004). I use three categories to describe these types of learning communities and note that they are not mutually exclusive. The three categories include (a) curricular learning communities, (b) student-type learning communities, and (c) living-learning communities.

Curricular Learning Communities

Curricular learning communities include linked courses or clusters where a cohort of students are simultaneously enrolled in two or more courses sharing some degree of explicit integration by way of curriculum, learning outcomes, shared assignments, and interrelated coursework. Team-taught courses are a more complex and labor-intensive form of curricular learning community because they require instructors to redesign the entire curriculum. Unlike course clusters, they are designed to include working with a large block of time that allows the faculty to "arrange and rearrange in class work to fit the goals of the program" (Goodsell-Love & Tokuno, 1999, p. 11). This model throws out the traditional format of an individual credit for each course and uses a mix of plenary sessions, collaborative projects, discussion groups, lab sessions, and multimedia to integrate interdisciplinary content around a specific theme (Lenning & Ebbers, 1999; Smith, 1991).

Student-Type Learning Communities

As the name suggests, this type of learning community is designed to cater to the needs of specific student populations. Learning communities may be for specific student groups, including students of color, honors students, students with specific academic interests, or students with disabilities (Goodsell-Love & Tokuno, 1999; Lenning & Ebbers, 1999; Levine, Smith, Tinto, & Gardner, 1999; Smith, 1991). This type of learning community may raise some questions about the effects of segregating certain groups in terms of the broader sense of community integration that learning communities strive to achieve. In response to this legitimate question, student-development theorists and researchers alike have argued that there can be great value in such groups (Astin, 1994, 1995; Chickering & Reisser, 1993). Often, students who have been marginalized in their educational journey need a safe, welcoming, and protected environment where they can become acquainted with collegiate expectations and simultaneously encouraged to add their own voices to the academic dialogue (Jehangir, 2002a). Also, it is important to note that heterogeneity exists within some student-type learning groups. For example, FG college students will likely include students from multiple ethnic backgrounds, cultures, and social classes. Likewise, a group of women of color is likely to reflect differences along the same dimensions. Note that student-type learning communities are developed with both the curricular and the social-involvement needs of specific groups in mind. They are, in many ways, an amalgamation of a curricular model with a student group. The fit of such models is largely dependent on the demographics, location, and student needs at each institution.

Living-Learning Communities

These learning communities, also referred to as living-learning centers, extend the learning-community concept to residence halls to overlap between the academic and social worlds of students. A residential college model is characterized by faculty residing in the same location as students. In this model, deliberately organized student cohorts register for a specific curriculum and live in a particular living space. The creation of the living-learning community therefore requires the coordination of not only the faculty but also other university systems, such as residential life and student affairs. The intent of these communities is not the return to *in loco parentis*. Rather, these residential communities, such as the Bradley Learning Community and the Chadbourne Residential College, both at the University of Wisconsin, seek to offer a thematic curriculum coupled with personal development and multicultural programs (Altschuler & Kramnick, 1999).

These three overarching approaches create a myriad of possibilities for redesigning space, time, and curricula to serve the needs of students and of particular institutions. The flexibility in this design lends itself well to developing diverse curricula and employing multidisciplinary approaches to engaging students in their learning. Lardner and Malnarich (2008) argue that "learning community work done well thus requires a skillful balancing of two moves: one structural, the other pedagogical and cross-disciplinary" (p. 32).

First-Generation Students and Learning Communities

Attention to the structure, content, and pedagogy that FG students engage in can serve to create a foundational experience (Engstrom & Tinto, 2008) for them, particularly in the critical first year of college. Learning communities present multiple possibilities for embodying the ideals of critical multicultural pedagogy for these students. The opportunity to develop relationships with faculty and peers is inherent to the design of learning communities, which cultivates a sense of belonging. This belonging can translate into increased engagement, accountability, and effort on the part of all student participants (Fogarty & Dunlap, 2003; Tinto, 1997; Zhao & Kuh, 2004). This type of curricular space can serve to draw students from the margins to the center, particularly when social integration extends into academic discourse allowing for deeper cross-disciplinary inquiry that builds bridges between the classroom and the personal experiences of students (Lardner, 2005; Tinto, 1997).

While the learning-community movement and other first-year experience programs are currently seen across campuses and institutional types in American higher education, many of the best examples of learning community impact on FG students come from community colleges.

These community colleges enroll 46 percent of today's undergraduates, and their student body encompasses a diversity far more reflective of the United States today than what is seen at most four-year institutions. Community colleges are the entry point for students with little or no "previous experience with higher education, low-income individuals and first-generation students" (Visher, Wathington, Richburg-Hayes, & Schneider, 2008, p. 5). This diversity in experience and background of these students includes race, ethnicity, culture, language, range of academic preparation, and life-work experiences. In an effort to serve the needs of these diverse students, learning communities are currently active components of the curriculum at 40 percent of community colleges around the nation (Visher et al., 2008, p. 5).

The burst of learning communities at community colleges was the result of a series of factors that dovetailed with the recommendations of the Commission on the Future of Higher Education (2006). These recommendations suggested that institutions focus on "access, affordability, the standards of quality in instructions and accountability to their constituents" (Visher et al., 2008, p. 1). While access to colleges, particularly community colleges, has increased significantly for FG students, the gap between academic preparation and college-level work, as well as other stressors, has resulted in high attrition rates and low transfer rates to four-year institutions (Engle & Tinto, 2008).

Efforts to bridge the gap between academic preparation and the demands of college-level work have historically been addressed by developmental courses. While some have long argued that these courses have no place in the college curriculum, Rose (2009) argues that if we "kick remediation off campus, the primary thing you will achieve is the greater exclusion of American youth from higher education" (p. 4). Research on the success of developmental courses has been somewhat unclear. Some studies demonstrate increased persistence for students who enrolled in developmental English and math courses (Bettinger & Long, 2005), while others demonstrate that while math remediation may lead to better grades in entry-level college courses, it had little impact on long-term persistence and degree attainment (Calcagno & Long, 2008; Martorell & McFarlin, 2007). In an effort to combat the uneven impact of past programmatic efforts, many community colleges have sought to redevelop developmental courses through supplemental instruction and learning communities that focus on discipline-specific help and opportunities to earn college credit toward degrees or certificate programs.

In addition, several institutions, both two- and four-year, have created learning communities to serve Title III (students) or Title

V students and have built successful learning communities to serve FG, LI and Hispanic students, respectively. In many cases, the focus of these group-specific learning communities has been to help students strengthen their academic performance, particularly in gateway courses such as math and science. Over 100 campuses have developed Emerging Scholars programs that link credit-bearing math and science courses to peer-assisted seminars that support students' transition from high school to college (Smith et al., 2004). At Tennessee State, a historically Black institution, the Emerging Scholars program integrates developmental courses with other university curricula in learning communities. This approach not only limits the stigma attached to developmental courses but also allows students enrolled in these courses to apply what they learn to curriculum to other linked courses (Carey, 2005).

As such, many colleges have revitalized their interest in learning communities as a model to build social and academic integration and support for their students. The Washington Center for Improving the Quality Undergraduate Education at Evergreen State College has also played a critical role in working with many institutions nationwide to develop meaningful learning communities, with a focus on educational equity that creates high-quality learning for "all students" (Lardner & Malnarich, 2008, p. 32).

What the Research Tells Us

A large body of research has focused on the effectiveness of different facets of the learning-community experience on students in higher education, at both two- and four-year institutions (Braxton, Sullivan, & Johnson, 1997; Smoke & Haas 1995; Tinto, 1998a; Tinto, Russo, & Kadel, 1994). Studies that have focused specifically on the impact of learning communities on FG college students have shown promising results with regard to social and academic engagement, as well as persistence from the first to second year of college. Vincent Tinto and colleagues began examining the impact of learning communities on nontraditional students at Seattle Community College, an urban, nonresidential campus catering to students with multiple life responsibilities.

Tinto, Goodsell-Love, and Russo (1993) compared the experiences of learning-community participants with non-learning-community students at the same institution. The results showed significant gains in persistence for learning-community participants when compared to similar non-learning-community peers. The study also suggested that curricular methodology incorporating peer groups and faculty creates "greater involvement in a range of academic and social activities and greater developmental gains than those in [the] regular curriculum" (Tinto et al., 1993, p. 20). In addition, students were more likely to take ownership of their learning when they saw that instructors were co-learners with them as they worked with different disciplinary material (Tinto et al., 1993, p. 19). Thus the multidisciplinary approach allowed students to have a diverse educational experience.

While the comparison study described above demonstrates that learning communities are beneficial along the dimensions of persistence, students' learning experiences, and their sense of belonging on campus, Tinto (1997) examined these potential benefits further by exploring "how the experience in the classroom matters" (p. 599). Using a longitudinal survey and qualitative case study methodology, his follow-up study examined whether learning-community participation at Seattle Central Community College made a difference for students' experiences, and if so, how that difference could be characterized. In addition to improved persistence in the two subsequent quarters, students also reported significantly more affirmative views of the college, the faculty, classes, and their own involvement in the collegiate community than their non-learning-community counterparts. These successes were attributed to "building supportive peer groups" (Tinto, 1997, p. 609), especially at a larger commuter institution.

In addition, students experienced "shared learning—bridging the academic-social divide" (Tinto, 1997, p. 610), where learning extended beyond the classroom and became a part of informal gatherings. Finally, students' qualitative responses indicated "gaining a voice in the construction of knowledge" (p. 611). Students seemed to value and appreciate the contrasting, yet complementary perspectives of the different faculty and peers. Students commented that the diverse learning environment allowed them to learn not only "*about each other* . . . but that diversity was an important factor in learning *about the content*" (Tinto, 1997, p. 612, emphasis in original). The outcomes of this longitudinal study thus lend support to the benefits of student involvement in learning communities, even in environments (commuter campuses) where building a sense of belonging and ownership is oftentimes challenging.

More recently, Scrivener et al. (2008) compared the experiences of 1,500 students at Kingsborough Community College. These students were randomly assigned into learning communities or unlinked courses. Students who participated in the 40 learning communities on that campus reported greater satisfaction with their college experience, a deeper institutional connection, a greater sense of belonging on campus, and that they were more engaged in their learning. These outcomes are mirrored in another multicampus study by Engstrom and Tinto (2007, 2008). These researchers used longitudinal surveys, qualitative case studies, and interviews to explore the effects of student behavior associated with engagement and persistence in relation to learning-community participation. This Lumina-funded study spanned 13 two-year colleges and six four-year institutions. The selection of community college sites included institutions where learning-community design "situated basic skill development within a broader academic context . . . and included low-income, first-generation and minority students" (Engstrom & Tinto, 2008, p. 8).

The results of this larger longitudinal study show that across institutions, students who participated in learning communities were more engaged in the classroom, in their relationships with peers and faculty, and they perceived greater encouragement and support on campus than similar students who were not enrolled in the learning communities. Moreover, learning-community participants were significantly more likely to persist from their first to their sophomore year than their non-learning-community peers. Of particular interest is that learning-community participation was "independently associated [with] persistence even after controlling for student demographics and engagement" (Engstrom & Tinto, 2008, p. 11). The authors of the study suggest that engagement alone cannot account for the increased persistence in participants; rather, there is something specific about the learning-community experience that sustains historically underserved students in their courses. The qualitative results of the study provide insight into the rest of the story and focus on the role of supportive environments, active and collaborative learning strategies, high expectations, and the impact of integrated curriculum and campus services on academic and social engagement and the persistence of underserved students.

While community colleges have been the focus of research on learning communities and FG students, changing demographics, financial pressures, and increased accountability surrounding undergraduate education have pushed research universities to explore learning communities as a viable means of engaging and sustaining all students on campus. The student body at many research institutions is changing, and diversity in race, ethnicity, class, and age creates new tensions and fragmentation. In the last 15 years, many research universities have began to seriously explore the benefits of learning communities with particular attention to the ways in which they can "make the scale of the research university more human" (O'Connor, 2003) and create more coherence in their undergraduate programs. While research universities have existing communities delineated by department and major, some liken these to "elitist communities resting on social relations that foster dominance and dependency, not the communities of inquiry needed in contemporary society" (Hill, in Smith et al., p. 50, 2004). Given the growing diversity at research universities and the low persistence and graduation rates of many students, especially LI students of color (Gerald & Haycock, 2006), the need to reexamine the role of curricular delivery and campus climate created inroads for learning communities. In some cases, learning communities were created to recruit and retain qualified students who opted to go to college out of state. The University of Maryland College Park created a series of living-learning communities around their general education requirements to support and engage students through their first two years of college (Shapiro, 2003). Temple University created their learning-community structures in 1993 with the specific intention of helping students develop a sense of community and improving teaching and learning for first-year students (Laufgraben, 2003). At the University of Michigan, learning communities have been a part of campus life for over 40 years and aim to provide their diverse student body with an experience that is similar to a small liberal arts college. Their offerings include 11 learning communities that range from residential learning communities to student-type learning communities to research-based communities and may include combinations of all these forms of learning communities. This hybrid design structure speaks to the diverse needs and interests of Michigan's student body, with the common goal to create meaningful relationships on campus and reinforce learning and student involvement both inside and outside the classroom (Schoem, 2003). These examples are indicative of what can be done on large, seemingly fragmented research universities to create meaningful learning experiences for all undergraduate students.

These innovations suggest that the time for reexamining how large institutions must respond to the needs of their many and varied constituents has arrived. While many research institutions have instituted learning communities to create a more coherent expression of their general education programs, there are some that have begun to dig deeper and consider how learning communities can be sites of access and excellence for students who have been previously absent from their campus communities.

In the following section, I will further highlight aspects of the learning-community design that are particularly salient to FG students. I will also further describe the ideals of critical multicultural pedagogy.

PUTTING IT ALL TOGETHER: BUILDING SITES OF TRANSFORMATIVE INTEGRATED LEARNING

During week one I mentioned I felt like I wasn't alone. I already felt like we were all part of each other. The closeness was not hard to find, one reason being that we had three classes together, but emotionally I felt connected.

—Diane, Asian American female student

Learning communities are by nature of their structure spaces that cultivate familiarity, and when done right, the familiar can be validating, reinforcing, inviting, and sustaining. It has been suggested that "learning communities are a purposeful attempt to create rich, challenging, and nurturing academic communities where they may not otherwise exist" (Gabelnick, MacGregor, Matthews & Smith, 1990). Given the invisible experiences of many FG students on large campuses, finding places where one is recognized and known by both peers and faculty can be powerful. Diane's comment speaks to the ways in which the structural design of learning communities nurtures connection.

Unlike classes that meet once or twice a week, learning communities with linked courses create opportunities for multiple encounters with peers and faculty in different contexts. As such, learning communities are changing the "basic organization of the curriculum to achieve more student-to-student interactions" (Barefoot, 2000, p. 15). These repeated opportunities for facilitated engagement can lead to a sense of trust and safety that is normalizing for many FG students (Engstrom & Tinto, 2008; Jehangir, 2008). Being validated and known in the classroom can also translate into increased confidence and engagement in the academic context, as well. It is in this process of creating an integrated community that we can best practice critical multicultural pedagogy.

While integrated learning is defined in many ways, I view it as the intentional process of bridge-building, where students and faculty make connections not only between disciplinary content and assignments but also between their lived experience and the curriculum. In addition, this process of integration involves a relational component—that is, when students make meaningful connections with their peers and their teachers, there is a willingness to think, write, and share dialogue with a degree of depth, trust, and engagement that is missing without intentional integration.

Another way to think about integration in the learning-community model is to consider the "degree to which [students] *incorporate integration across courses* and the *complexity of collaboration involved*" (Goodsell-Love & Tokuno, 1999, p. 13, emphasis in original). *Integration* implies the extent to which the courses themselves are connected, and *collaboration* implies the level of joint activity of both students and faculty. Research on learning communities has highlighted the ways in which integrated social and academic engagement, or blurring the lines between these two ways of being, can result in higher-order critical thinking and a deeper understanding of academic material (Barefoot, 2000; Engstrom & Tinto, 2008; Stefanou & Salisbury-Glennon, 2002; Tinto, 1997; Tinto et al, 1993). Tina, a White FG student, wrote about ways in which cross-disciplinary connections and sharing assignments within her learning community (also referred to as *dovetailing*), impact her learning:

> The three classes go together so good, it is almost scary to me. The material discussed in each relates to each other so well that I almost get the feeling that it is planned that way. I find that to be very helpful because when I am doing an assignment, especially a paper, I can refer to the other materials in each class for references. I also find it helpful because I am starting to feel knowledgeable on the issues that we are touching on. Each topic that we fit talks either about race, class, gender, or multicultural experiences in each of the three courses. This really affects my learning experience because if they did not go together, I probably would not understand the material as much because I would not get to have the opportunity to be taught from different teaching perspectives and see others views.

Tina's comment addresses the ways in which integrated learning can occur when curricular linkages extend beyond common themes to explore the connections, conflicts, and gray areas among diverse disciplines and narratives. The core practices of integration can support deeper thinking and can set the stage for higher expectations of students and how they might begin to view themselves as thinkers and knowers (Domizi, 2008; Engstrom & Tinto, 2008).

While diversity and active learning are core practices within learning communities, it is the extent to which these practices contextualize the curriculum, assignments, and projects in the lived experience of student participants that creates critical multicultural spaces. bell hooks (1994) argues that "theory is not inherently healing, liberatory or revolutionary. It fulfills this function only when we ask that it to do so and direct our theorizing toward this end" (p. 61). As such, not all learning communities will be critical spaces for students. Learning communities do, however, have many of the ingredients essential to create classrooms that take a multicultural, dialogical approach to teaching and learning. By itself, diversity is simply a descriptor of participants or curricula. Yet when diverse students actively engage in diverse curriculum as listeners, storytellers, and problem solvers in the community, a shift occurs where they are no longer outside of the learning experience. but active contributors to it.

Problem posing is one approach that critical pedagogues often employ in working with students. The act of problem posing extends beyond sharing personal narrative and includes the practice of situating problems in one's experience, listening and investigating a problem from different lenses and coming together to solve it (Wink, 2000). Problem posing involves "inductive questioning strategy [where] students can ground their discussion in personal experience and integrate that experience into the broad social context" (Wallerstein, 1987, p. 27). This practice in critical pedagogy connects well with learning-community reform movements, where knowledge is viewed as relational and connected and where the interdisciplinary curriculum is built around themes and problems. Lauren, a multiracial student, wrote about how examining racism from multidisciplinary perspectives, lived experience, and the narratives of her peers allowed her to come to new understandings about economic and social inequities: "My learning experience within the community over the last 15 weeks has been evolving even up to this very moment. Though our classes are no longer in session our finals week still allows me to reflect and put into perspective my whole experience. I remember feeling like I was witnessing a new way of learning about society and history. This was not traditional and though it was taking facts (even the horrible ones) and examining the behavior of man. As we addressed economic domination, I remember balancing the learning with what I was experiencing in my life today."

Many institutions around the country have deepened and broadened their commitment to diverse students by capitalizing on multicultural, interdisciplinary curriculum in learning communities. LaGuardia Community College and the University of Michigan, two very different institutions, have developed learning communities focused on intergroup dialogue, service learning, and curriculum built around racism, sexism, and heterosexism that encourage students to embed their cultural capital into the classroom and think together about addressing problems of inequity (Mellow, van Slyck, & Eynon, 2003; Schoem & Pasque, 2003). Thus learning communities are well suited as sites for inclusive curricula and pedagogy. The flexibility of their design can allow instructors to refine the structural dimensions to add components ranging from extracurricular activities to advising partnerships and to residential living.

Intentional Partnerships

Higher education scholar Clark Kerr wrote that "the multiversity is an inconsistent institution. It is not one community but several . . . devoted to equality of opportunity, it is itself a class society. A community . . . should have common interests; in the multiversity, they are varied, even conflicting. A community should have a soul, a single animating function; the multiversity has several" (1995, p. 14). Kerr's comments highlight the ways in which university life can be a disconnected meritocracy for not only faculty and staff, but especially for FG students who are trying to make sense of their many new environs.

As campuses seek to support students, particularly at large, public four-year institutions, these students are often simultaneously encouraged, invited, or drafted to participate in numerous programs, both academic and co-curricular. They may receive grants that require a certain amount of community service or check-ins with program advisors who are distinctly different from the academic advisors they must meet with in order to register for classes. Rose, an African American student, captured the fragmented nature of her campus experience, even with regard to programs explicitly created to help support her:

> One of my multicultural scholarship programs is trying to come up [with] some way to keep us connected. You get a scholarship after you get admitted into college [and] it was kind of like, "OK, we will check up on you around midsemester and after the semester." But there wasn't really any connection; I could walk past a bunch of people and not have no clue that they were in the same program, and of course the next year's generation I really didn't know who they were. So when they come to these meetings, well, we started having these mass gatherings and it was like, "I have seen you before, but I had no clue you were in the program," "I had no clue, you know, I had a class with you. I didn't even know you were in this scholarship program."

Rose captures the lack of coherence in her collegiate experience. Her dilemma is reflective of what many students experience, particularly on large campuses. While learning communities alone cannot bring together fractured communities on campus, they can provide structure and can serve as a model for connecting students, faculty, and staff with each other to *practice* community. Intentional partnerships start within the teaching and learning circle of a learning community, where teachers come together to discuss and support the success of each student. This can be as simple as requiring student conferences that allow students and teachers to connect one-on-one outside of the classroom and focus on particular academic or personal interests and areas of growth.

In addition, many learning-community models have considered other campus parties who are critical stakeholders and advocates for students in their learning communities. These stakeholders may be academic advisors, staff from cultural centers, or liaisons from service-learning centers who become extensions of the learning community by way of presenting some information in class, attending and facilitating activities outside of the classroom, or triaging with faculty to problem solve specific challenges that might arise during the semester. Residential living-learning communities, where faculty or advisors live with students or are involved in residential hall programming, provide multiple opportunities to engage with their students in out of class and build social networks with students (Inkelas, Daver, Vogt, & Leonard, 2007).

While Inkelas et al. (2007) demonstrate the positive impact of living-learning communities on FG students, it may not be feasible for many FG students to reside on campus and benefit from such opportunities. Other means of bridging the divide between academic and student affairs may be to have one advisor assigned to a given learning community and build in-classroom activities that include registration planning, financial aid support, or use exploratory tools such as StrengthsQuest or the Myers Briggs Type Inventory to allow students to consider how their interests connect with their academic pursuits (Schroeder, Minor, & Tarkow, 1999). Other models have incorporated first-year seminars into academic learning communities with the intent of supporting both intellectual and social transition into college. These seminars can be taught by advisors, instructors, or advanced peers (Henscheid, 2004).

Ananda, an African American student, spoke about the extent to which these early and intentional learning partnerships supported her well beyond her first year of college:

> It's helped me, because, early on, I was able to build a network of friends, and of staff. Not only the learning community teachers but also with my advisor. And other people who were involved with the TRIO program. That just helped out a lot. 'Cause I know a lot of people who

> came in freshman year, they don't really know anybody, but I had about 30 core friends who I saw every day. And of course all the staff was willing to do whatever they could to help you out. Even today, if I still want to, I can go to my first-year advisor and still ask for advice and tips. If I'm having trouble with my current advisor, I can't get in contact with him, I can ask my old learning community advisor whatever and he's going to help me out. So I think that's been very, very helpful, and again. . . . I think just the network that I've built has helped the most and has stuck out for me the most.

In many ways, learning communities by nature of their design push us to consider the student experience holistically and ask us—staff, students, and faculty alike—to engage in the care of our students and of one another. In doing so, we cannot simply retreat to the confines of our disciplinary homes, department, or private classrooms. Learning in the community demands that we are, in a sense, like our students: exposed, challenged, and surprised. This process pushes us to practice unlearning and relearning and being in dialogue within ourselves and with our students to transform how we learn together.

In order for faculty, staff, and students to come together in transformative ways, universities need to reexamine the attention, design, and value of undergraduate education on their campuses. This chapter opened with the words of architect Christopher Alexander (1977), who addressed the importance of intentionality and the universality of purpose in designing meaningful and functional spaces for multiple constituencies. Higher education scholar Clark Kerr (1991) echoed the critical need for universities to return to the drawing board to consider how to design connected, inclusive communities to serve undergraduate students. Learning communities are one model of intentional and integrated design that many institutions have considered in reforming undergraduate education on their campuses.

Using the ideas presented in this chapter as a springboard, the next chapter will the provide the rationale, design, and challenges involved in creating a Multicultural Learning Community (MLC) for FG students at a large predominantly White public university. The MLC is not presented as an ideal, but rather as a demonstration of the practice of multicultural critical pedagogy within a learning community that serves FG, LI students.

Part III

Getting Through

Lessons from First-Generation Students

Chapter 5

Rationale and Design for the Multicultural Learning Community

Many of my doubts and fears are gone. I have learned just as much from the students as I have from the teachers. I realized that we are all in this together and that we have to help each other in order to succeed. College is not about the one who is the smartest but more so about the one who is willing to stay until the end. College is not about being alone. There are some people who realize that college is something we have to do together, and those are the people I want to surround myself with. This is all due to the learning community that I realize this. I think that all first year students should have to enter in a learning community. I feel this was the best way to leave high school and enter college. I could not have made it [without] the learning community, who are now more like family. To everyone I say thank you.

—Sasha, African American female student

In the quote above, Sasha reflected on her experience in the Multicultural Learning Community (MLC) at the end of her first semester in college. She highlighted the ways in which learning together facilitated her transition to college and invited her peers to be teachers and co-learners in her educational journey. She also alluded to challenges in the journey, noting that it is not merely "book smarts," but the capacity to persist that is critical to completing college. To provide context to the experience and voice of the student participants, this chapter provides an overview of the learning-community design, the rationale that shaped the curriculum and pedagogy, and how and where it was programmatically situated on campus.

The MLC included three college courses: a first-year writing composition course, a creativity art lab humanities course, and a multicultural relations social science course or an international literature course. The learning community required concurrent registration in all the courses, and enrollment was held to a maximum of 24 students. These learning-community courses were offered at a freshman-admitting college at a large public university between fall 2001 and spring 2007. Enrollment in the MLC was managed by the TRIO Student Support Services (SSS) program housed in that college. Like most SSS programs, this one offered counseling and support to its students; however, it was unique in the extent to which academic engagement was deeply embedded into the programmatic offerings. Working with faculty from the college, this TRIO SSS program had established a series of credit-bearing learning communities and supplemental instruction courses. The MLC was one of nine learning-community and supplemental-instruction registration options offered to incoming students each semester, and it was the only learning community that included concurrent registration in three courses.

In spring 2001, my colleagues Patrick Bruch, Patricia James, and I came together to design a learning community for FG students. Each of us had previously taught several courses that enrolled a critical mass of FG students, and we were eager to move to a deeper, more integrated model of teaching and learning. In addition, each of us had already established connections with the TRIO Student Services program in our college, and we wanted to create a learning community in partnership with student services that would allow for intentional support and communication between the curricular and co-curricular aspects of the students' experience.

The goals of our MLC fell into two categories that were closely linked to each other. The first category pertained to programmatic goals and involved assisting students with the transition to college. The transition process incorporated academic and social support systems into the learning-community design and actively provided structured "footholds" to develop critical thinking, writing, and verbal skills in the learning experience. The second category reflected the curricular goals of creating a learning environment that fostered collaboration, creative conflict, and knowledge construction.

A central component of building academic and social integration was to build intentional partnerships between academic and student affairs. While our learning community was designed by faculty, we collaborated regularly with advisors in the TRIO programs. During the first two years of the program, one of the MLC instructors also served as an academic advisor for the students in the program. When staffing changes required a shift, a TRIO advisor was assigned to the MLC, and faculty and advisors communicated regularly in person and via email regarding student progress, challenges, and successes. This partnership allowed us to understand the progress of individual students more holistically and provide timely support or referrals when necessary.

The main purpose of the MLC was to create a space that offered a curricular structure that would enable FG college students—many of whom were also students of color—to explore issues that engaged them and would help them create a place of belonging in the academy, given their marginalized status in a predominately White institution. In order to create this sense of space and place, the MLC design incorporated a small class environment, linked course offerings, and coordinated curricula among the three learning-community courses. The faculty instructors for each course worked together to create thematic links in their curricula with the goal that students would examine issues of identity, community, and agency, each of which ties to issues of race, class, gender, disability, and sexual orientation, from a multidisciplinary, multicultural perspective. These linkages were designed to allow students to apply diverse interdisciplinary theories of multiculturalism to their lived experiences, while simultaneously empowering them to find their own voices as a means of knowledge construction within the context of a shared learning experience with other participants. The faculty also worked together to develop collective curricular goals to improve students' critical skills in the areas of writing, creativity, humanities, and social science as multiple ways of knowing (Jehangir, 2008). *Multiple ways of knowing* is a phrase best associated with the work of Howard Gardner (1983, 1998) who broadened the confines of what constitutes intelligence by arguing that in addition to logistic or mathematical intelligence, human beings possess the capacity to exercise linguistic, spatial, kinesthetic, musical, intrapersonal, and interpersonal intelligences. The MLC sought to invite students to recognize and understand these different strengths they brought to the classroom and also encouraged them to think of multiple ways of knowing as different lenses from which to identify, assess, and evaluate the issues we were studying in the classroom and in context of their lived experience. One approach was to consider how the different disciplinary areas provided frameworks to explore problems and how using multiple frameworks to consider issues of identity or prejudice deepened one's understanding of these issues. Students were also encouraged to use different modes of expression to present, design, write, or create assignments.

Three themes shaped the curricular and pedagogical design of this learning community: the themes of identity, community, and agency, which were derived from the theoretical framework of critical and multicultural pedagogy. *Identity* refers to examining the self, the identities of others, and exploring identity in sociopolitical and historical contexts, media, and art forms. Examining the self requires reflection and the development of skills to position and reposition one's self in the context of multiple identities and issues raised in the MLC. *Community* refers to the active development of place and belonging within the learning-community courses. Developing community requires considering students' perceptions of what it means to be a community and inviting them to contribute, critique, and examine forces that impede as well as enhance the growth of the community. Moreover, developing community asks students to engage in constructing knowledge, building relationships, and taking ownership of the learning processes. Finally, *agency* refers to the process by which engagement in a learning community and around multicultural issues encourages and empowers students to examine issues of social change and civic engagement in the context of their own experiences and the experiences of others and apply these new understandings to advocacy for social change.

Each of these themes extend critical multicultural pedagogy by asking students to engage in the process of examining the self, to consider the self and others in the context of community, and to explore ways in which newly constructed knowledge emerging from this dialogue can be applied to impact social change. Most importantly, an examination of these issues is situated around the deconstruction of power and invites students to consider the influence of intersections of race, class, and gender and how power dynamics play out to the advantage of some groups over others (Jehangir, 2009b).

Demographics of Students in the MLC

Between fall 2001 and spring 2007, the MLC was offered seven times. Over the course of these seven cohorts, 130 students participated in the MLC. Females outnumbered males overall by almost two to one: 84 females (65 percent) to 44 males (34 percent). Race broke down across cohorts as follows: 42.9 percent African American or Black, 29.7 percent Asian American, 10.9 percent White, 8.6 percent Hispanic, 5.4 percent Native American, and three students listing no racial data. The average high school rank for all cohorts was in the 56th percentile and the average *composite* ACT score was 18.

BRIEF HISTORY OF GENERAL COLLEGE

The MLC was housed in and taught by faculty in the General College (GC) at the University of Minnesota. At the time, the freshmen-admitting GC offered general education courses designed with attention to development and multicultural curricula. A key function and mission of the college was to provide a means of access to the state's flagship research institutions for students who would otherwise be denied admission (Higbee & Lundell, 2005). Brubacher and Rudy (1997) state that "it is mainly due to the influence of state universities (and related institutions such as land-grant colleges) that an ever larger proportion of the total American population was getting a chance at higher education" (p. 160). GC was born out of this sentiment.

Founded in 1932, it was one of two innovative programs at the University of Minnesota "initiated by President Lotus D. Coffman to provide a more general education to a broader range of students that was possible under the prevailing liberal arts curriculum" (Johnson, 2005; University of Minnesota, General College, 1997). Originally a degree-granting institution, GC later became an access point for students who did not meet the admissions criteria for other degree-granting colleges. GC students enrolled in challenging, college-level coursework and completed general requirements in the social sciences, humanities, and sciences before transferring to their major of choice. Given, however, its unique access mission, particularly within a research university campus, GC courses were often perceived as less rigorous by many in the university community.

In addition to innovative teaching and smaller class size, GC students participated in intensive career and academic advising programs. Fall 2004 figures indicated that GC enrollment was approximately 1,800 students, of whom 875 were first-year students. Forty-eight percent were students of color and 51 percent were female (University of Minnesota, 2004). GC also housed programs such as the Commanding English program, which catered to students who were English language learners, and the TRIO Student Support Services program, a federally funded program that serves low-income (LI), first-generation (FG) students, as well as student with disabilities.

In fall 2005, as part of a large university strategic repositioning plan, GC and two other colleges were closed. Most of the faculty and staff in GC remained together in a new department titled Postsecondary Teaching and Learning, which was merged with the College of Education and Human Development. As a result of the merger, the College of Education and Human Development launched a freshman-admitting program, and the department of Postsecondary Teaching and Learning currently offers an interdisciplinary, multicultural First-Year Experience program to all 400 first-year students admitted to the college. All students participate in the first-year program, which includes an interdisciplinary team-taught first-year inquiry course and learning communities. Of the over 400 admits to the college, 250 are students who have been historically underrepresented in higher education.

BRIEF HISTORY OF TRIO PROGRAMS

All students enrolled in the MLC were participants in the TRIO SSS program. TRIO programs were first funded under the Higher Education Act of 1965 with the intent of creating more opportunities for educational access for lower- and middle-income families (Groutt, 2003). The first program under the TRIO moniker was Upward Bound, a demonstration project that sought to "identify secondary school students from low-income backgrounds who were underachieving and to motivate and prepare them to pursue postsecondary education" Groutt, 2003, p. 2). The 1968 Higher Education Amendments led to the availability of grant funds to create SSS programs (Groutt, 2003).

The TRIO SSS program, housed in GC and now in the College of Education and Human Development, has been funded since 1976. It has been designated an "exemplary site" by the U.S. Education Department's most intensive study of nationwide TRIO SSS program performance. During the summer orientation program, the TRIO SSS program invites approximately 240 eligible students per year to participate in a TRIO SSS program. The invited students are those who are perceived to be in greatest need of support during their transition to college. In addition to LI, FG status, factors considered when discerning which students are fit for the program include: "students with the lower precollege preparation, older students, students of color, student parents, and students with disabilities" (Schelske & Schelske, 2002, p. 8). Students who opted to participate in the TRIO SSS program at GC were assigned a TRIO advisor and participated in an "intensive intrusive academic advising program based on a sophisticated and comprehensive student academic progress monitoring system" (Schelske & Schelske, 2002, p.40). In addition to individual advising in areas ranging from financial aid and personal counseling to study skills development and career planning, the program offered career and educational planning workshops, along with leadership and community-based events. These academic and educational interventions are "proactive and intense and tied to the academic arena in which students operate" (Schelske & Schelske, 2002, p. vii). These interventions include a series of course offerings with supplemental instruction seminars and learning communities, of which the MLC was one option. With the closing of the General College, the TRIO SSS program is currently housed in the College of Education and Human Development and the a new iteration of the Commanding English program called College English Transitions programs function under the umbrella of the TRIO program.

TEACHING TOGETHER AND APART

Teaching has historically been a private activity enacted in a public space. That is, the instructor uses the course requirements as a guideline for curriculum selection, but the ultimate choice of what and how to teach lies with that instructor. In a learning community, even one in which an instructor retains control over his or her individual class, the act of linking with other courses demands an openness to talk and think about not only how instructors teach but also how they know. When my colleagues and I came together to create a learning community for FG students, we knew we held common values about the role of education as an avenue to justice and personal development, but we did not know exactly how each of us enacted these values in our teaching.

In the MLC, my colleague Patricia James taught the Creativity Art Lab course, Patrick Bruch taught composition, and I taught a social science course titled Multicultural Relations. In the final two iterations of the MLC, I taught a postcolonial international literature course. Although the courses were thematically linked together, each instructor approached this linkage from the domain of his or her discipline. As such, each course had its own purpose and objectives. It is helpful to consider how each instructor shaped his or her course with specific regard to its purpose, objectives, and the nature of the assignments. In examining these interdisciplinary starting points, we can better understand the thematic and dynamic interplay of borrowing and sharing perspectives and content across the courses.

The Creativity Art Lab invited students to "use the arts to better understand our own experiences and gain a deeper understanding of what it means to be human" (James, 2002). The focus of the course was on "the arts as a way to explore, express, and critique themes of identity, multiculturalism, and social justice" (James, 2002, p. 1). Using masks, text, spoken word, drama, dance, and music, students engaged in a process-oriented classroom experience. Central to this classroom experience was the opportunity for students to consider how art helps them better understand their world, express experiences within it, and work for change. Students were reminded that this process involves risk taking, exploring unknown dimensions of thinking, and inventing creative ways to complete assignments.

In the Writing Lab, students were asked to engage in a process of "helping each other strengthen [their] abilities to participate in and reflect on the processes of academic writing" (Bruch, 2002, p. 1). Topics for the lab focused on "large public conversations that create truths about places (like cities), events (like the Los Angeles uprising), and issues (like crime or education)" (Bruch, 2002). At the heart of these units of reading and writing are questions of "who benefits in what ways from different versions of the truth, and questions of how writing participates in truth making" (Bruch, 2002, p. 1). The primary objective of this course was to provide students with the structure and support for developing a skill set that would allow them to write critically, incorporate their own experiences into their writing repertoire, revise and adjust their writing for specific audiences, and conduct efficient research using multiple sources. The Writing Lab's approach to the themes of identity, community, and agency was embedded in the "perspective that writing is an important way of making (or at least shaping) things like truth and relationships among people" (Bruch, 2002, p. 2). As with the Creativity Art Lab, this course acknowledged that writing, like performance art, involves personal expression and that "our writing locates us in relations with others" and "involves a lot of risk" (Bruch, 2002, p. 2).

The Multicultural Relations course, and alternately, the International Literature course sought to examine historical and contemporary multicultural relationships within American and global societies. Using interdisciplinary material from history, sociology, psychology, personal narrative, and literature, the courses both examined the social construction of classism, racism, sexism, and other inequities in various contexts. The central goals of these courses included examining the theoretical perspectives of systems of oppression and engaging in a dialogue about if and how these theoretical perspectives are reflected in students' daily lived experience. Students were asked to consider the power of language and history and how it continues to impact and shape current social and institutional inequities. They were also encouraged to bring their lived experience into the classroom and explore connections between theory and current societal practice.

When Bruch, James, and I came together to plan this learning community, our conversations on the surface were about what we could teach, but the subtext was much more about how we taught and what we valued in teaching and learning. In many ways, our faculty learning community mirrored the student experience in that we searched for common touchstones but also laid claim to pedagogies that defined our teaching. Like the students, we had to learn to trust one another and open the practice of our teaching to critique, development, and growth. In our earliest conversations, each of us drew on our disciplinary and theoretical frameworks to come up with themes for our community. Critical pedagogy (Kanpol & McLaren, 1995) multiculturalism, social learning theory (Bruffee, 1999), and constructive developmental theory (Baxter Magolda, 1992, 1999; Kegan, 1994) all informed our collective practice and informed our belief that FG students needed more opportunities for personal and academic learning in their critical first year of college. As each of us was deeply interested in what students brought to this experiment in the learning community, we gave attention to the cultural capital of our students and how that could inform and interface with our curriculum.

The key commonalities that enabled us to value each other's contributions were our shared beliefs that participants are co-learners and that the key to learning is reflective participation. We settled on community, identity, and agency as organizing themes for the learning community because these concepts legitimized each of our theories of education, gave credence to students' experiences, and invited them into challenging conversations (James, Bruch, & Jehangir, 2006, p. 11).

As teachers, we engaged in plenty of challenging conversations ourselves—in planning the curriculum, we began to negotiate many aspects of teaching together. Everything from a common attendance policy to how we enacted these community themes was part of the dialogue. Once the learning community was under way, we met weekly and compared the progress of the students in each of our classes, worked on strengthening the links between assignments and discussions, or found ourselves adding new threads to the curriculum based on something that occurred in class that week. We often debriefed one another about our time in the classroom, shared frustrations about things that did not go as planned, and regrouped after a class or when a new assignment fell flat.

Through this process, we were constantly deconstructing our own teaching, questioning, articulating, and refining the purpose of certain assignments, as well as encouraging and supporting one another. We found that we began by speaking in our own disciplinary voices, but over time, we engaged in one another's disciplinary language and began incorporating new ideas into our teaching. James, the art faculty member, often talked about metaphor rather than themes, and I found myself incorporating metaphors into my teaching more and more. This process of borrowing and sharing concepts, and in some cases, redefining our approach to teaching, pushed us to become co-learners along with our students. It caused us to let go of the old familiar ways we each had come to think about teaching and pressed us to take risks we might have never opted to take as solo teachers.

LEARNING IN COMMUNITY

Each of the three courses in the learning community addressed the themes of identity, community, and social agency. At several points in the semester, the instructors created assignments, activities, and projects that asked students to think about the intersection of these themes within and across classes. The curricula for all three classes included theoretical articles, first-person narratives, and art (visual, dance, theatrical, and musical) that explored issues of racism, classism, sexism, and other inequities. Students were consistently asked not only to consider this material in the context of their own experiences but also to consider experiences, perspectives, and positions that were different from their own. Several times in the semester, all instructors and students met together in one classroom and engaged in a reflective activity around these themes. For example, in the first week of class, both students and faculty discussed the meaning and history of their names. Sasha, an African American student, reflected on the activity:

> The activity that made me feel welcome was when we talked about the history of our names. This was something that I had never done before. I didn't even know the history of my name, but listening to everyone else's history gives me a sense of who they were. Many people had stories behind their names. Their names represented something that was important in their lives. And even those of us who didn't know the history of our names wanted to learn it after that. Now I want to know more about my own name. This made me feel welcome because I felt important. I felt that anyone who wanted to know about my name must want to know about me. So for me that was an excellent icebreaker.

This name exercise illustrates one way in which students' life experiences and identity were brought to the center of the learning community. These types of active learning pedagogies were also steps toward creating inclusion and encouraging students to think about their own self-authorship in the context of learning. As instructors, this relational component of the course extended beyond social engagement; it also shaped and deepened academic engagement. As students came to see one another more holistically, challenging discussions around race, class, and gender were no longer dialogues among strangers, but rather conversations among individual students and their histories, their narratives.

Students also participated in numerous collaborative projects throughout the semester, including formal group presentations, peer reviews of writing, leading discussions, and creativity exercises. Each semester culminated in a capstone project that centered on the art class but used concepts and themes from all three courses. The capstones included public group performances or large scale murals that were displayed in public places in the college (Jehangir, 2009a). An active learning pedagogy was central to our project, and collaboration and interdependence were critical components of many assignments.

The act of engaging in collaboration and working together moved in stages, and as is well known in the development of group dynamics, our process was not linear. To help facilitate an environment that could tolerate the vulnerability, discomfort, and ambiguity of exploring complex ideas with diverse participants, we always started each semester with a "social contract" (Smith, MacGregor, Matthews, & Gabelnick, 2004, p. 102)—a mission statement that provided some parameters for how we might engage. We divided the process into two parts: first we asked students to reflect on the characteristics of strong versus weak teams. Working in groups, students came up with lists of behaviors and expectations that they shared with the collective. Students were asked to back up with their lists with stories that helped us better understand what they meant by "respect" or "trust" or "rudeness." In the second phase, we drew on the ideas of each group to coauthor a statement that would reflect how we hoped to be together as a community. At critical points in the semester, we revisited these ground rules to remind ourselves of how we hoped to *learn community*.

INTERDISCIPLINARY SOCIAL AND ACADEMIC INTEGRATION

As a learning community, we also attended some activities away from campus together. The outings varied but typically included attending a theatrical performance or visiting an art gallery. We sought out plays and art exhibits that shared concepts or themes with the material we were studying in our classes. This off-campus event also included sharing a meal together. Students often talked about this event as one of the most memorable in the semester. The reasons were varied—for some students the act of engaging with peers and faculty away from the classroom increased their comfort level and sense of belonging with the group. Mike, an African American student, wrote,

> I felt most engaged the day that me and my class of students went to see a play called Con'Flama. This play was very interesting, especially the set up and props. I feel that my class enjoyed it overall because it was a great experience. It was the first time we as a class ever [were] together outside of class—we were all having fun laughing, joking, conversating, and so on. We all enjoyed ourselves that night we even had all of the teachers from our learning community attend. It actually felt sort of like a day with my family. Like a family outing that consisted of a group of friends that I felt like I knew forever. It was late night I thought no one would show but everyone did. We all spent time as a class together learning and understanding this play. I kind of gained a little knowledge from my classmates of the definition [of] a community.

For some students, these exhibits and activities reinforced the multicultural curriculum and reflected the culture of our participants. Ariel, a Chicana student, described her experience on one such outing: "Going to the art gallery was very overwhelming for me. Being a Chicana myself, I was happy to see my people express them and do it so well, making master pieces of art. So many art pieces touched me in many ways. Seeing religion portrayed a lot in the paintings I wasn't surprised to see. Seeing all the art made me proud. In ways it inspired me to be more creative and show my views on Chicanos and their ways of life, dreams, past."

For others, attending a performance or viewing art that reflected or connected with their lived experience was powerful—it became an extension of the curriculum and reinforced their understanding of texts and narratives addressed in the classroom. Kelly, a Native American student, wrote,

> These classes helped me see and enjoy the play in that I understood it. It seemed to relate to all the classes in our learning community. Without these studies I doubt I would have been able to relate understand and connect the three classes to the play in a way that I would appreciate it. Readings from each class, the discussions of the Watts riot and talking about art body movements, all connected to the poetry in the play. . . . All classes stimulate different parts of my mind. Helps me think better; in different views and approaches to life, home, school, work, expectations of my community.

In our efforts to create a coherent, integrated, and interdisciplinary space, we built formal and informal relationships among our three classes around three dimensions: community development, texts, and themes. The three themes of identity, community, and agency have already been described above. Regular activities, such as the name game and developing a class mission statement or attending co-curricular activities, focused on creating spaces for students and faculty to see each other as co-learners and deepen the relationship between our collective ability to reflect on our identities within and distinct from the classroom. By using our classroom texts (written works, films, art), we also created activities or assignments that invited students to think about intersections and divergent ideas that emerged from each class. For example, during the course of one week, the theme of identity was explored from the position of three different texts. In one class, students read from Omni and Winant's article "Racial Formations" (2001) and explored how social, political, and economic forces shape the perceptions of race and racism, and the students discussed their own racial identities. In the composition course, the readings focused on representations of youth in 1960s America. In the art class, students explored the concepts of the *private*, *public*, and *unknown self* to consider which aspects of the self we chose to reveal in given contexts. Discussions across the three classes focused on multiple identities—a topic particularly salient to our FG students.

During another week, students were asked to express identity in less traditional ways. Working with a photomontage assignment and metaphoric "I am" writing (James, 2000), students used images and texts to express their identities. Juxtaposing images cut from magazines, students were asked to create a photomontage of some aspect of their identity. They also wrote corresponding "I am" paragraphs that helped them interpret the metaphors and meanings of their own work. Rita, an Asian immigrant student, created a photomontage in which "the image shows a sensuous arm reaching across space to celebrate the transition from girlhood to womanhood" (James, 2000). Her corresponding "I am" statement read,

> I am coming out of innocence
> Entering through life,
> Passing by the mysteries,
> Full of certainty,
> Carrying along hope,
> Hiding the secret fear,
> I hope to make it through,
> I hope to break it through,
> Until I reach peace and love for eternity.

As students looked at one another's photomontages, they became more aware of their peers' intellect, creativity, multiple identities, and vulnerabilities: "Multiculturalism was a lived reality, not only an abstract idea" (James, 2000). Most of the students had never made a photomontage before, so James provided examples, guidance, and set up parameters; however, she also left room for students to figure out what they wanted to do to fulfill the assignment. In this creative process, they made meaning not only about their identities and the creative process but also about their own thinking. As Rita wrote: "Photomontages never made sense to me because they were just the mixture of too many confusing things, until I made one myself. It's really fun; it makes you think in an unusual manner. From bottom to top and west to east, just awkwardly. It is surely fun but not easy. You really have to think!"

By making and talking about their photomontages, students also developed a deeper understanding of the social construction of knowledge. This learning took place through their senses, bodies, and emotions, not only through facts or theories. Students gave voice to realities not traditionally honored in higher education. In doing so, they were embodying ideals of critical multicultural pedagogy by drawing on the cultural capital of the students and bridging the processes involved in personal and academic development (James, 2005).

Diversity and Reflection

Given the multicultural demographics of our class, students were encouraged to think about diversity and multiple identities in the context of disciplinary perspectives, their own life experiences, and from the narratives of their peers. Making sense of multiple ways of knowing pressed students to let go of the idea that there is only one right answer, and they were encouraged to grapple with shades of gray, both cognitively, emotionally, and interpersonally. This process often gave rise to dissonance, caused by grappling cognitively with new concepts or challenging social issues and dealing with internal disequilibrium. While the process of engaging with this diversity was not easy, many students often came to value the questions and answers posed by diverse contexts, peers, and disciplines. At times, this created "a-ha" moments that were validating and reinforcing. Shawna, an African American student, discussed how the racial diversity of the classroom, together with a sense of community, impacted her learning. She wrote, "Especially now in college I can say that I'm way more aware of races outside of my own. Many times in life, we have the tendency to lean toward what we know. This semester we were able to be in a comfortable climate where we can work outside of a comfort zone and really get to know one another without the feelings for awkwardness or being forced into something with other people. I'm a lot more open not that I wasn't before but now in a more mature way."

In addition to race and ethnicity, many of the FG students connected around issues of social class. Arrianna, an African American student, recalled a moment when one of her peers addressed challenges with poverty. She wrote, "In the fourth week, our class made a big communication accomplishment. Rachel broke down the barrier between people and their feelings in the class. She made me and others in the class see her side of the story—of growing up poor."

Yet these positive insights or epiphanies did not always come easily; in fact, many students truly grappled with the way in which the curriculum, their peers, and the instructors challenged their previously held beliefs about a range of topics, including perceptions of the self and others. Mallory, a biracial student, wrote about the challenges of having a discussion within a diverse group about the history of slavery:

> Well, it was a long week as far as actions. I just want other people to listen to me like I listened to them. If I ever spoke up after someone said something it was because they weren't giving me good reasons for their comments. It was Sasha who just wasn't willing to hear what I had to say. For me under the context of that question saying slavery is wrong was not an answer and it was just their way of simplifying a much more complex question. I am not so sure that they understood where I was in my statement.

Given these heated moments, opportunities to engage in reflection were critical to all participants in the community. Weekly written reflections were required in two of the nonwriting intensive learning-community courses. These reflections allowed students and faculty to engage in low-stakes dialogue with one another about experiences in the classroom and allowed students to step back from specific moments to digest what had occurred. Mallory, the same biracial student mentioned previously, addressed the role of reflection in both the cognitive and emotional aspects of her learning. She wrote,

> After collecting all of my learning logs, putting them in order, and then separating them into sections, I started to read them and discovered how useful the learning logs are for this project. Not only did it make me remember some of the subjects under discussion more clearly, but reading the logs brought back some of my other thoughts and feelings at the time. When I thought about some of the feelings I had not expressed in the logs I realized why they had not been written in the logs. At the time I thought they might be unnecessary, harm my grade somehow, or harm other people's view of me. I know now that

> the reason they were not in the logs was because they truly are hurtful opinions to others. They are now hurtful even to me. This class has taught me a lot and so have the people in it. The logs have been a big help in keeping my memory intact and unbiased, which is most important because I think we only want to remember the best in ourselves.

Some questions for weekly reflections were written by faculty and drawn from Stephan Brookfield's (1995) work on *Becoming a Critically Reflective Teacher*. Other reflections were particularly student directed and did not include any questions as prompts. As instructors, the reflections allowed us to talk with students who were particularly challenged by various discussions and topics and gave us insights into both the strengths and weaknesses of our classroom dynamics. The reflections also became a site of private dialogue between an individual student and a faculty member and facilitated trust that allowed both parties to engage with each other with a degree of vulnerability and compassion that would not have been accessible without these reflections.

ENTERING THE UNKNOWN

Teachers who strive to craft critical, multicultural classrooms "deliberately involve themselves in messy but crucially important problems and when asked to describe their methods of inquiry, they speak of experience, trial and error, intuition, and muddling through." (Schön, 1991, p. 43)

To bring any group of people to together to explore issues of identity, community, and agency is to embark on an exciting and challenging journey. Each individual brings his or her history, narrative, and belief system to the table, which can both advance and hinder the progress of the community as a whole. When we consider that many first-year college students are but three months removed from high school and are at the formative stages of their own adult development with varying degrees of maturity, the potential for the unexpected to occur is high. In our particular community, FG students were also trying to balance the new role of college student with other work, family, and life expectations. As such, there were many occasions when emotions ran high, students reverted to adolescent behaviors, or created cliques that were detrimental to the larger community.

Some learning-community practitioners use the term "hyperbonding" (Smith et al., 2004, p. 102) to refer to behaviors of subgroups of students within learning communities who lower the bar by forming exclusionary factions, engaging in rude behavior, or doing little work. Research by Jaffee, Carle, Phillips, and Paltoo (2008) point out that these negative social dynamics are the unintended and often underdiscussed aspects of learning communities. In some cases, strong primary group relationships that develop in peer cohorts can put students at odds with faculty, as well (Jaffee, 2007). Indeed, versions of these concerns played out to varying degrees in our MLC, and again, we drew on ground rules, open conversation, and student reflection to address and redirect these issues. In many ways, these challenges were part and parcel of inviting students to navigate the rather ambiguous territory of an interdisciplinary, critical, and multicultural space. Shani, an African immigrant student, commented on what she viewed as inappropriate classroom behavior:

> I really was disgusted with today's discussion. I honestly felt like people weren't respecting their classmates or their teacher. People were talking when others were; they were talking without raising their hand, interrupting, and disregarding the group that was up there. I felt really sorry for the group presenting because the class was up in an uproar while they were supposed to be teaching us. I understand the importance of being engaged in a discussion, but today, people crossed that line. It wasn't just discussion, it was plain and simple rudeness. I want to say yes that our classroom community is working, with the exception of today's class. Today's class was not an example of a functional community because people only wanted to talk, and not listen to others. Listening is a HUGE part of being in a community and I think we all need to be reminded [of] this sometimes.

Rosa, a Chicana student, expressed frustration with the childish behavior of two students in her working group. She wrote, "Well, this week I thought that our group was not interacting like a group should. I feel as if it really doesn't feel like a group yet. For one, Lavon and Shawna keep arguing about how she doesn't like Lavon talking and how he doesn't like her looking at him. I just feel that they argue about the wrong reasons. I think that they should just deal with the fact that they are both in the same group and that they are going to have to work together whether they like it or not."

In both cases, the behavior of a small group of students negatively impacted the learning experience of their peers. This discord gave students an opportunity to name and determine what they expected of peers in their community. Students also came to understand and value the role of conflict and accountability within a community. As the teaching team grew more experienced with working in the community, we were able to stem potential "hyperbonding," and oftentimes, we shared examples of problematic group dynamics with students at the beginning of the semester when students were developing their social contract. We shared with students specific examples of behaviors that, while initially playful or innocent, had far-reaching negative consequences for the collective. But frequently, the difficult moments were just that—difficult, uncomfortable, and tiring. As Jaffee and colleagues (2008) have suggested, these types of classroom dynamics could shift the focus off thoughtful academic work. I argue, however, that facilitating these sometimes inevitable group dynamics is part of building community, and the payoff in terms of social and intellectual engagement far exceeds the inevitable conflicts of working with others.

In our learning community, the students were most likely to test the boundaries between themselves and instructors, as well as among peers, when working on our final cumulative project. The final project was either a large public mural or a collaborative theatrical performance. The final project sought to draw on material from each class in the learning community and embody the themes covered during the course of the semester. Texts, writing, and other collective work from the semester could be incorporated into the project or drawn on for inspiration. Even after spending weeks together and having a sense of ownership of their learning, the process of actually creating a collective final project was not easy for students. For one thing, many students had come to a new understanding of themselves and their voices, so the final project became a training ground to lay claim to their ideas, beliefs, and identity.

As they became more adroit at dealing with ambiguity, students were willing to push one another more and take less direction from the instructors. Working to guide and facilitate these larger projects meant that we had to engage in the real practice of critical multicultural pedagogy. As faculty, there was a sense of urgency for students to begin to actually *do* the work of bringing together a finished product. At the same time, the debates, discussions, and disagreements the students engaged in were the very processes of meaning making that instructors hoped they would they would utilize during the project. While the process was unscripted, heated, and nonlinear, it also represented the extent to which each student cared about and was invested in the final product.

Khodija, an African immigrant student, wrote about how challenged and frustrated she felt in the planning stages for a mural: "The last few weeks have been very frustrating. I never thought that it could be so hard. Planning the mural caused a lot of arguments and people just getting against each other. It was a very long process and it took a lot of yelling and arguing to figure out the theme for the mural. Every time we came up with a plan someone had to come against it and that made it harder on everyone. I really thought that we will not pull this off."

Students were very vocal about their frustration with the process of coming together to create a large project. They expressed concerns about their own ability to contribute something meaningful to the task at hand. Some became disenchanted with the long, drawn-out discussions and were disengaged for a period of time. Others spoke only when an issue was critical to them, choosing to stay out of the fray for other parts of the discussion. Some tried to facilitate compromise. Despite their individual position or stance, most students wanted their lived experience, their identities, and their ideas to be represented. Particularly when working on a mural that had some permanence and posterity, students—the same students who had described being invisible in high school—now wanted to leave their mark on the campus. In their reflections, many students addressed how this project extended beyond the end product and was an exercise in learning to work in collaboration with diverse ideas, people, and contexts. Jennifer, an African immigrant student, wrote,

> Besides art I have learned a lot about myself. I learned that in order to work well with groups I have to listen with an open mind and most of all try to consider their thoughts. While making this mural there were many arguments and disagreements but throughout the whole thing I did not realize that it was part of participating in this mural project. I have written about it in my reflections by saying "I thought there was no point in today's class because all we did was argue." But you replied by saying "maybe this is part of the process." I tried to look at it your way by listening to my classmates and understanding what they have to say. After the mural was done I thought to myself that you were right—it was part of the process. I have learned that in order to accomplish a big goal like making a mural there are ups and downs and people are not going to agree with each other all the time but the outcome makes it all worth it.

As such, the learning community was an exercise in embracing the possibilities of classrooms as democratic multicultural communities. The process was not always pretty, but for many of our FG students, it served to provide a foundation whereby they could think about themselves as co-constructers of knowledge. While we worked to acculturate students to see themselves as part of the university, we also aspired to create a space where they could practice acculturating the university to make way for their cultural capital and many selves. To this end, the final phase of this project ended with follow-up interviews with MLC students, three to four years after their enrollment in the learning community, to explore what long-term impact, if any, this experience had on their college trajectory. The next sections give context to the interviews and themes.

MOVING FORWARD: LIFE AFTER THE LEARNING COMMUNITY

As stated in the introduction, the second part of this study investigated if and how the MLC experience challenged the isolation and marginalization of FG students beyond their first-year, and if their first-year participation had any long-term impact on the rest of their university experience. To explore this question, students were recruited by using a purposeful sample. Juniors, seniors, or recent graduates who had participated in the MLC between the years of 2001 and 2007 were contacted via mail and email and were invited to participate in the interviews. Twenty-four students responded to the invitation and were interviewed when they were juniors, seniors, or recent graduates. Interview questions covered four specific areas: students' MLC experiences, their university experiences outside of the MLC, involvement in extracurricular activities, and their future goals.

Of the 24 students who participated in the follow-up interviews, the majority were students of color. Fourteen students were African American or Black (of which four were of East African descent), three were Asian, two were biracial, one was Hispanic, and five were White. Forty percent were male and 60 percent were female. Of the 24 participants, 23 were still enrolled at the university or had recently graduated, and only one had dropped out. The absence of more voices from students who did not persist at the university is an obvious limitation of the study.

A team of three individuals—two graduate research assistants and I—examined each of the interviews using narrative and case study data analysis procedures. Our analysis consisted of three phases. In the first phase, each researcher individually engaged in a process of meaning making by creating cases for eight randomly selected interview transcripts. In the second phase, we began by focusing on "emergent construction" (Denzin & Lincoln, 1998, p. 3) and focused on categories that emerged in one interviewee's text, cross-checking these emerging categories with the other participants' texts (Flick, 1998). In our notes during the data readings, we paid specific attention to "categorical aggregation" (Stake, 1995), or the extent to which certain ideas reoccur in the students' interviews.

This process led to an emergence of new themes and categories as the data analysis continued until there was a "saturation of categories" (Lincoln & Guba, 1985, p. 350). Our final phase of analysis was data reduction, which included "simplifying, abstracting, and coding" (Miles & Huberman, 1994, p. 10) the data in accordance with the established categories or themes. The coding involved "identifying information about the data and interpretive constructs related to analysis" (Merriam,

1998, p. 164). Although 10 themes emerged from the data analysis, I will use the remaining chapters of the book to focus on the three overarching themes, which are (a) belonging, (b) academic and self-identity, and (c) critiques of the academy. Drawing on the extended interview narratives of a select group of seven MLC interviews, the remaining three chapters will animate these themes through student voices to consider how these themes might inform our practice in higher education.

The decision to highlight the voices of seven students in particular is not to discount the other students; instead, it is a way to allow readers to engage in-depth with a sample of students and to understand their experiences across themes. The intent of this approach is to showcase these narratives and provide a deeper, richer picture of the experiences of these FG students in college. The voices of the remaining 17 interview participants are represented throughout the book. The seven students whom I will introduce below represent all four cohorts of the interview participants. Each was selected because they reflected the diversity of the learning-community participants across race, age, gender, immigrant status, language, academic interest, and varying ranges of persistence at the university. Some of their names and stories may be familiar because earlier chapters included excerpts from their narrative writing *during* their first year of college, while still enrolled in the MLC. The following chapters, however, focus solely on their post-MLC interview data to provide a window into how these students reflected on their collegiate journey three to four years after the MLC experience.

BRIEF INTRODUCTION TO THE SEVEN FEATURED STUDENTS

Ruben

Ruben is an African American student who came to the university from a midsized midwestern city. He is the youngest of his siblings and the first to attend college at age 18. Even in his younger years, his family always looked to him as a problem solver and caretaker. Reflecting on how he felt about coming to college, he said, "It was never really was for *me*—it was always with the understanding that I'm the one who is supposed to do this, which added to the stress." At the time of the interview, he had graduated from college with a self-designed degree in psychology, sociology, and youth studies and was pursuing a graduate degree out of state. He noted that finally now this graduate degree was "for him" and felt like he had given himself permission to claim his educational journey. During his undergraduate years, he was involved in the Black Student Union, Alpha Phi Alpha fraternity (a historically Black fraternity), and he participated in the National Student Exchange program. He was also employed on campus at both the university admissions office and the National Student Exchange office, where he focused on working with multicultural populations. Since the interview, he has completed his master's degree in psychology and was accepted into a doctoral program.

Anna

Anna is a Latina, FG student who came to the university directly after high school. At age 18, she often cared for her younger siblings and assisted her family members with daycare. She worked both part-time and full-time jobs while enrolled as a full-time student in college. She recalls withdrawing from some classes during a particularly trying semester and being placed on academic probation and suspended. She appealed the suspension and was able to return to complete her degree. When asked about involvement in campus activities, she noted that she "worked a lot and just didn't have time for it," but she did volunteer with her neighborhood Cinco de Mayo program, local Boys and Girls Club, and through her church. She majored in family social science, a multidisciplinary major that to prepared students to work with families in a range of professional environments, and she graduated six years after enrolling at the university.

Jarod

Jarod arrived at the University at age 23, five years after graduating from high school. An FG college student, Jarod was employed off-campus for about 25 hours a week. He was encouraged to apply to college by his younger sibling, who was already enrolled at the university. An avid reader, Jarod was especially interested in history and music. When describing his motivation to come to college after several years in the workforce, he said, "This might be little naïve, but I went to college basically just to learn. I felt like before I went to college there was something missing and I would read books on various topics and my brother would be like, 'you know you can go to school and get credit for doing that?'" A self-described introvert, Jarod was deeply interested in social justice issues and was proud of his Italian American heritage. He studied Italian and participated in a study abroad program in Italy during his junior year. He majored in political science and graduated five years after enrolling at the university.

Lauren

Lauren is a multiracial student who was 26 when she participated in the MLC. She describes her own culture as a blend of Native American, Puerto Rican, Creole, and Black. She first enrolled at the university about five years prior to her MLC experience and said she "didn't do too well [in her] first year, dropped out, got married and returned back after many years." As an adult, married student, she wondered how she would fit into to college this second time around and found that she was comforted by her FG peers and "completely fell in love with her studies." The learning-community focus on social justice and art sparked her interest in women and art and led to a three-week study abroad program in Europe. Following the program, she chose not to return to the university and also got divorced. At the time of the interview, she continued to express interest in returning to higher education, but she noted that "her loans are in default so she can't go back to school right now." Lauren moved to a large midwestern city and created a nonprofit organization for transitional housing for the homeless.

Zahara

Zahara came to college when she was 20 years old. Having been out of high school for close to two years, she did not view herself as a traditional-aged college student. In her own words, "Since I came in as an older student, it was a little harder to get connected." She worked throughout her college career, assisting her mother with a family business, as well as well as working other jobs in retail. She also cared for her young siblings. Reflecting on her time at the university, she said, "I didn't really do much on campus, but off-campus, I am a family girl." She had two younger siblings who went to college; one of them graduated and went on to law school. Zahara identified herself as African American but noted that her family line included White, Black, and multiracial people and joked that her family referred to themselves as the "All Yellow Posse in reference to the light-skinned people in [her] family." Zahara majored in family social science and graduated six years after enrolling at the university.

Paul

Paul enrolled at the university directly after completing high school in the United States. He richly described himself as "a Liberian, an African American, Black, Christian, a meditating person, a frat boy, a soccer player, and a student." Following his participation in the learning community, Paul, an FG college student and immigrant, became involved in many campus organizations. While working about 12 hours a week, he was a residence hall advisor, a member of the Black Student Union and Alpha Phi (a historically Black fraternity), and he had joined the campus ROTC program. When asked about his many commitments, he said, "I just want to be part of something where I can grow." Double majoring in sociology and youth studies, Paul hopes to go to graduate school one day. He is currently a sixth-year senior and noted that he took some time off school "because [he] had to take care of [his] family back home."

Davu

Davu matriculated to the university after graduating from high school at age 18. He came to America from Ethiopia at age nine, leaving behind his parents and siblings to live with his uncle and cousins. When reflecting on his childhood, he said, "I grew up quick; at 14 I had my first job; at 17, I got my first apartment." During his college career, Davu worked full time, often taking night jobs as a parking garage attendant to support himself. A self-described history buff, Davu is passionate about equities issues in the United States and Africa. Double majoring in political science and African American studies, Davu has been an active and involved participant in his Ethiopian community since high school. In college, he became president of the Ethiopian Student Association and brought middle school and high school students of Ethiopian descent to campus in order to connect them with future educational opportunities. In the future, he hopes to go to graduate school and to work for the United Nations, Red Cross, or an organization that can make a difference in the lives of others. Davu is currently a sixth-year senior, primarily because and took one semester off college to return to Ethiopia for his brother's graduation from medical school.

The last three chapters of this book focus on the narratives of these seven students in the context of the three metathemes that emerged from their interview data. Beginning with the theme of belonging, I will draw on their narratives to unpack the strengths and limits of critical multicultural pedagogy and learning-community design in challenging the isolation and marginalization of FG college students. These student voices provide context, raise questions, and offer critiques of learning communities and higher education as an idea, a space, and a community.

CHAPTER 6

BELONGING AND FINDING PLACE

No house should ever be on a hill or on anything. It should be of the hill. Belonging to it. Hill and house should live together each the happier for the other.

—Frank Lloyd Wright, American architect

Belonging is a complicated dance of giving and receiving. To belong is not only to be welcomed but also to feel that we have something to contribute. To belong is to find connection, to leave our mark and to be understood and valued for our unique contributions. Like the house and hill, connectedness to people or a place or an institution is enriched when all parties are critical to the making of that relationship. When first-generation (FG) students were interviewed about how they came to belong in college, they were acutely aware of occasions when they were welcomed but not on their own terms. They were adroit at distinguishing between when they had established authentic connections with peers and when they were pretending to fit in. They were aware that belonging can be elusive, just as when we shift our place of residence or leave our familiar hill, we must start again to try to find home.

This chapter addresses the theme of belonging and place that emerged from the interviews with the participants of the Multicultural Learning Community (MLC). The interviews discussed in this chapter focus on the seven students who were introduced at the end of Chapter 5. The chapter begins with students' reflections about their first days as college students and moves to their recollections of how the MLC experience provided support, navigation, and familiarity that grew into belonging. The chapter concludes with students' reflections

...t it means to gain and lose belonging and how they sought out new validating spaces beyond the learning community.

Part of understanding what it means to have a sense of belonging is to know what it is to be alone. For many FG students, their need to belong to a community stems from an understanding of what it feels like to be isolated or lost. Many FG students have strong familial and community connections before they come to college. For them, the loneliness on campus is heightened by the stark contrast between the community-oriented culture to which they may be accustomed and the more individualistic culture of college.

Many students feel lonely or homesick when they initially come to college; for FG students, however, this experience of isolation is layered and multifaceted. Some feel invisible as a student of color on a predominantly White campus or question their academic educational preparation and language proficiency. Other FG students spend limited time on campus due to employment or family roles. All of these obstacles add up to make the adjustment to college challenging. Students captured this feeling of isolation when they described their first encounters with college education. For some students, this experience rested in a fear of the unknown or of not knowing what to expect as FG college students. For others, it was fear of being academically marginalized because of perceived concerns about their academic preparation, test scores, and ability to succeed in college classes. A number of students questioned if and where they were going to fit in and be accepted—what would their "place" look like?

Given this problem, students said that needing acceptance was a critical component in developing a sense of belonging within a class or even an institution. Many MLC students talked about belonging and finding place as a process that grew out of their interactions with peers, faculty, staff, and a curriculum that created opportunities for personal and intellectual engagement. For each student, different aspects of this process and different sources of support became salient to their experience.

TESTING THE WATERS

Finding place is a theme that characterized the experience of the MLC students, who were able to discover a space of belonging, both literally and figuratively. It was not something that happened immediately but rather grew and developed over the course of the semester. In fact, students' earlier recollections of college reflected their doubts and even distrust of faculty and peers. Students began to slowly test the waters and dip their toes into this idea of *learning as community*. As students spoke about their experiences, they illustrated how interactions and social and academic engagement helped them cultivate a sense of belonging. However, they also reflected on the anxiety they experienced as they came to college. As Jarod thought back to the first day of school, he recalled that he "was scared to death, basically." Five years out of high school at age 23, Jarod found himself considering his age, his less-than-stellar high school experience, and his FG status, and he questioned how he would fit into this new experience at the university. Lauren, another adult student, was 26 at the time of her participation in the learning community. She recalled her feelings about her first days in the community: "In the beginning, I was excited because I thought it was a cool idea to have all the classes with the same people. But I was kind of nervous, because I was older; like, most of the others students are fresh out of high school. I was returning back to college after a long time, and I was nervous, because I was wondering how was I going to connect? How was I going to be able to socialize? And how I would feel on a daily basis being around a younger generation?"

While the older students questioned how they would fit in with a crop of younger peers and worried about their rusty academic skills, traditional-aged students, such as Anna, were equally overwhelmed by the size of the campus and questioned their choice to attend a large university. Anna remembered her fears at the beginning of the semester: "I was so intimidated; I mean, I knew that I wanted to go to a bigger school, but at this university, there are three campuses! I am alone, but I mean, I knew I had to build relationships with everyone else." Like Anna, each student experienced a sense of being alone and apart from others, and they questioned how they would get a foothold in this first year of college. As the weeks progressed, these students each spent close to nine hours a week together in class with their cohorts, and their feelings of isolation began to diminish. In each class, students were encouraged to share stories, opinions, and ideas about themselves, their cultures, and their identity. Paul remembered that his initial reaction to the learning community was skepticism. He recalled having strong misgivings about an educational environment that wanted and expected students to share their perspectives and experiences, but over time, he began to appreciate the value in building connections with peers and faculty:

> At first I didn't even understand why is it we had to talk about ourselves out here, but then I learned to appreciate the fact that we did that. Because when you walk into a class and someone knows your names—your first

> name and your last name—it means something. In university here you are walking around campus nobody really knows you; I mean, there are thousands of people here, but you entered into the [MLC] community you didn't have to explain yourself, and the more people know you the more you build a bond over the course of time.

Paul's comment alludes to the ways in which the structure of learning communities can facilitate students to learn in relationships with one another. The process by which students arrived at a place where they did not have to "explain" themselves is explored further in the following section.

TO KNOW AND BE KNOWN

Students reflected on how their time in the learning community allowed them to know and be known—that is, they began to establish relationships of mutual understanding with others that allowed them to reveal their many selves. Students spoke of building a social network of peers whose life stories and educational experiences created points of connection among them, despite differences in age, race, gender, or ethnicity. Although these points of connection were not without for contention or disagreement, students generally expressed a feeling of trust, care, and accountability for one another.

For some students, relationships developed in the class became the core of their social network in their first year of college. Paul recalled this process of developing friendships:

> At first it started out pretty slow, trying to get to know each other and just building that relationship. Everything was different; I mean, it was new, but as time went on, I kept seeing the same faces in the same class—like, I saw them maybe three times a week, who knows, maybe six hours a day . . . so really it was like we had a relationship growing up as a group. There was this whole sense of awareness of everybody, and we had lunch together. I used to have some of my MLC friends come in the dorm and eat and come up to my room and we'd just kick back and listen to music until it was time to come to class. We did our homework or reading together, so it was like there was this sense of community, and people just kind of connected.

Like Paul, Ruben remembered the co-curricular activities that grew out of connections established in the classroom and took on a life of their own. Students began to form meaningful friendships. Although these relationships had their ups and downs, students had a sense of commitment to one another. In some cases, these relationships extended well beyond the first year of college. Ruben recalled how these co-curricular activities solidified relationships with his peers:

> I remember we went to a play downtown—the first time I had been anywhere outside of school because I am not from Minneapolis and that sparked off everything. Once we were taken outside of the classroom, it was like we figured out that we all have these things in common and from there we became, like—in a positive way—a clique. And we started arranging our own, [Laughs] little parties, and celebrating Dr. Martin Luther King's birthday together, we had a potluck, and it was just arranged by us. We were very, very attuned to each other and what was going on even though we didn't like each other at times. You know, it was almost like your brother or sister at times, but you still care for them a lot.

Anna echoed the feeling of being family—she recalled moments of frustration and irritation with her peer group but also a sense of being sustained by them:

> I kind of feel like we were *one*. Of course there were conflicts and disagreements and arguments. And even through all of that and through people getting mad at each other, people being crabby, I mean, us getting sick of each other, we all were there for one reason and that was to learn and we all stuck through. We all had the same thing in mind and it was to complete this [degree]. I think that was special and that was definitely meaningful, even though there was tough times in that community we all stuck through it and were one—we were one.

Parks (2000) contends that "we need a place or places of dependable connection, where we have a keen sense of the familiar: ways of knowing and being that anchor us in a secure sense of belonging and social connection" (p. 90). As Anna suggested, places of "dependable connection" are not without contention, nor are they idyllic; what distinguishes these places from others is the capacity to feel valued and secure, especially when other aspects of one's life might be less stable.

Davu spoke about the way in which the learning community created a sense of condensed campus, a more manageable way to navigate a large campus and also commented on the practicality of having a peer group who shares in a collective weekly schedule and classes. He said, "I am very close to some of the MLC students. I think we see the importance of this learning community, and in a sense it was three classes into one. It's a very different scenario when you come to college and actually have the high school atmosphere within the college.

Because in high school we see everybody in the same class and it was just like that. We were more attached to each other. Because people understood my weekly homework, she understood, he understood my weekly homework and when I would have free time."

For students like Zahara, this immersion experience with her peers became a safe haven, a sort of oasis from the challenges occurring in her life. She recalled one particular semester that was a difficult time in her personal life, but her connections with peers and faculty in the community allowed her to be intellectually engaged without having to hide her emotional challenges:

> I was going through a tough time, and my high school sweetheart had gotten in trouble and had gotten indicted and was going to prison. And I remember through the learning community, it helped me cope, I think, because the class was so fun, first off—all the people were great, and then the group of people was just comical. Like, we were learning, but we were being *us*. I will never forget that semester—it helped me cope with that situation. He would be gone for eight years. Everything was going on, but because of this community, I was able to get through the semester with good grades. It was helpful because I was emotionally kind of attached to the community.

The act of telling others about her boyfriend's incarceration was an indication of Zahara's comfort with the MLC community, and it also reinforced her comment about learning while continuing to be "us." She did not expect her problems to be solved but appreciated that she did not have to disguise what was going on in her life; rather, her lived experience could be embedded in her learning.

Although emotional connection, and in many cases, developing friendships extended well beyond the classroom, for some students the comfort and trust developed within the boundaries of the MLC itself was the defining feature of belonging. The opportunity to spend large amounts of time together in their LC classes helped foster positive social interactions for students. Each student translated this social connection into different aspects of his or her collegiate experience. Jarod gained a sense of acceptance and validation from his peers, which was new to him. Prior to this experience, he may have chosen to remain anonymous in a classroom setting, but the format of the learning community slowly allowed him to gain a comfort level with his peers:

> Because we were always around each other. People were kind of forming friendships and things. I feel that—I don't know how to say it—the sun and the moon were in alignment. Everything seemed to work for the MLC. We were down in the lounge. Just always,

> like, sitting together. Like those kinds of things stick out. I just really remember being really nervous at first. And just kind of getting myself comfortable with getting to know people. Just opening up . . . I guess. I was so used to, like, a work environment. You go, punch in, and do your thing. And, like, college was just, like, a really foreign thing to me.

For many students, this connection with the students in the MLC became their primary source of social engagement on campus. The learning community participants became their core group of friends, their network, and their comfort zone. Jarod described these strong social connections, and for him, these relationships translated into a support system for testing his ideas and grappling with the complex issues covered in the curriculum: "I mean, my first year of school those were the only people that I really hung out with. Again, just being in a group that really backed me up on a lot of things. I mean, people like Jermaine and Ruben, they really, like, validated my ideas, and I think about the fact that I did so well . . . grade-wise . . . in those classes . . . all of them. It was really easy for me to see that I can do it. I guess, just that it was possible [Laughs]. So after that first semester I was like, 'This college thing is a piece of cake.'"

Lauren, another adult student, had experiences that coincide with Jarod's interpretation of belonging. She acknowledged that given her age, she was less available or even interested in some of the extracurricular activities that her peers engaged in. For her, the value of being understood and gaining a richer understanding of her peers created opportunities for more authentic and deeper intellectual engagement:

> I liked the fact that it was the same people. Yeah, I felt comfortable with the other students, regardless of the age. 'Cause like I said, you really formed . . . not so much a relationship, but more of a comfort level that you won't expect, especially whether you're returning back to college after a long time or you're a freshman. For me, it made me more comfortable with my own intellect, I guess. You feel comfortable, like you're around people that are willing to be open and share. Because a lot of times you have class, with, like, what—100 or 200 people? But because it was more small and more . . . I don't want to say more intimate because it's corny. But you feel more involved rather than robotic. The social experience, I think, for me . . . because I was older, and everyone knew I was older . . . it stayed in the classroom. It stayed kind of more or less still connected with our studies. I remember there was times where people were talking about parties and what they're going to do over the weekend and normal stuff.

Jarod and Lauren addressed how gaining a comfort level and sense of belonging with their peers allowed them to feel safe enough to share their ideas. Their comments demonstrate how a sense of belonging allowed them to gradually contribute more to the academic dialogue. This process of contributing to classroom discourse was an empowering experience for them as students and demonstrates how a sense of belonging can serve as a bridge between social and academic development. This reaffirms the value of supportive peer groups (Tinto, 1998b) in creating integrative learning experiences, particularly for historically underserved groups.

Navigating the University: Faculty, Staff, and Student Interdependence

Three to four years after the MLC, students recalled the roles that peers, faculty, and staff played in sustaining them throughout their first year, providing a network of support as they transitioned out of their first year and into new experiences in college. These recollections, of course, might suggest that this support was a one-way street, where students came to teachers and advisors and received some nuggets of wisdom or kind words that helped them on their way. I would argue, however, that what sustained these relationships was the degree of interdependence that developed among faculty, students, and staff as co-learners in the community. Instructors relied on advisors for insights and ideas that may not have been visible to us from our vantage point, and we all relied on the students to challenge and teach us, as we hoped to do the same for them. The students captured the delicate balance of our collective relationships in their comments about the MLC. For example, Ruben described his interpretation of the faculty-student connection: "The best thing about it all is it wasn't a typical teacher-student thing—we didn't have inappropriate relationships, but it was truly a community. Those presentations that we did where all three of the faculty were present. . . . That's how tight we were; the material we covered went in depth, it was that togetherness, and we still have that now. Once we were done being classmates and learning from one another, what can we do for each other now? I would still like to know who you are. That is so powerful."

Anna reflected on her close relationships with her teachers and advisor. For her, the close relationship was not only about support but also about feeling "special" and cared for:

> I remember especially my advisor—I remember feeling a different kind of connection with her. You know, it was more . . . I don't know . . . not

> on such a professional level. I mean, it was still professional, but I think that I warmed up to her a little better because it was more personal and I felt comfortable coming to her with issues. And I remember the writing professor, he was so hilarious and he was just an easy guy to talk to, you know, and I think they put a lot of time and effort into teaching us, because we were a community and because I just feel like we were a special group, you know, and it made everything special.

Ruben also discussed how having faculty and peers as advocates redefined how he felt on campus. He argued that building these mentoring relationships was instrumental in claiming a place on campus. Describing how he felt while attending a co-curricular activity with the MLC, he recalled how "we all stood by each other and we could see that we were important . . . we knew we were important people." This experience of feeling "special" and "important" was an expression of the students' sense of validation and underscores that "a mentoring community can confirm faith that they will be a new home" (Parks, 2000, p. 93) especially for FG students who often feel displaced on campus.

Both Ruben and Anna also speak on the issue of blurry boundaries, where the relationships between faculty and students are not "typical." They hastened to add that there was nothing inappropriate at play but that there was a feeling of shared community. Parker Palmer (1998) considers this complex interplay of roles as teachers, students, and advisors seek to develop a learning community. Given that power is always an issue in the classroom, Palmer argues that in meaningful communities, power differences in the context of race or gender can be challenged head on, yet often the difference in the power status of the student and teacher can be stickier to balance. Teachers might thrive on giving students space and ownership of the classroom, but there is a need for the teacher to maintain boundaries that sustain a healthy community. Then there is always the issue of grading and the power it wields over students. Palmer (1998) writes, "The real threat to community in the classroom is not power and status differences between students and teachers but the lack of interdependence that those differences encourage. Students are dependent on teachers for grades—but what are teachers dependent on students for? If we cannot answer that question with something as real to us as grades are to students, community will not happen. When we are not dependent on each other community cannot exist" (p. 139).

This effort to develop interdependence requires a considerable amount of risk on the part of all stakeholders. For teachers, it pushes us outside the comfort zone of our material and disciplines and into the complicated world of our students' lived experience. While we are not all trained counselors, and it is not in our capacity to solve all of their problems, a willingness to know and acknowledge what sustains and challenges students is essential to our collective progress. Lauren reflected on the value of being able to talk about her challenges with her instructors: "They were very supportive and understanding about my personal issues. Going back to college and then being married—there was a lot of upheaval in my home, and so I think that the fact that the professors were accessible. They weren't stand-offish, they wanted to understand you. They made me feel like, I had support."

Like other students, Jarod did not expect that his instructors could solve his problems, but he did value being able to talk openly about his fears as an FG college student and appreciated knowing that there were individuals who could consistently serve as signposts as he maneuvered through university life. He said,

> My relationship with my LC instructor really helped. Back then, the first year of college, I don't think I would have done as well. She helped me a lot with understanding how to, I guess, maneuver through academia. . . . I think just in general the whole experience really . . . gave me a lot of confidence. I feel that if I hadn't done that I don't know if I would have continued college. I mean, after a year or two, I could have seen myself dropping out. Because, again, I came into college really apprehensive and not knowing what the hell I was doing, I guess. And not knowing why I was here, and if it was going to work, and all these things. The teachers were really receptive and really like nurturing, I guess. It really helped me feel confident about myself as a person but also as a student.

For many students, meaningful connections with MLC faculty and advisors also translated into a sense of accountability to themselves as students and to their peers in the community. Paul said,

> Honestly, the MLC instructors, I look up to them. They were the foundation of my college life—they and Nolan, my advisor. I really look up to those four. It is not like I need someone saying you can do it, you can do it. But I would feel embarrassed if I was in one of the LC Instructors' class and I didn't turn in my work. I am serious, man—that's how bad it is. If you have that kind of relationship with teachers, it's like, she knows you can do better. And if you are out there clowning [Laughs] it's like—I am serious—like, I mean, I just can't look her in the eyes.

Zahara's comment picks up on this idea of accountability and expresses an awareness of how her instructors walked a fine line between encouragement and high expectations. She understood that the outcome of her grades and her learning were her responsibility:

> Our LC instructor was so compassionate. She listens to you, you know what I mean?—she's very attentive, and she's not, like, What do you want today, and what's your issue today? She actually wants to know what's wrong. It was important for us to understand the work and it was important for us to turn in our papers. It wasn't because *she* wanted us to do it, but was important for *us*. She made that clear—that it's not about me, it is nothing about me, it is all about you. You know what I mean? This is your grade, and you take responsibility for what you need to do. She was on top of us, but she still was compassionate, she still cared. I think that is the hard thing that she had to do, is be this caring person and then be, you know, the person who is supposed to lay down the law.

This theme of supportive interdependence as central to belonging reinforces the value of paying attention to both the inner and outer aspects of learning—emotional and cognitive, subjective and reason based (Rendón, 2009). While awareness of both these facets of learning is salient to all students, it has particular relevance to FG students whose previous educational experiences may have reinforced the disconnect between who they are and how they learn. Rebuilding this connection is a critical pathway for them to cultivate a place in the academy. Kuh, Kinzie, Schuh, and Whitt's (2005) study of effective educational practices on college campuses demonstrates that high expectations coupled with meaningful and collaborative engagement between faculty and students affirm that underserved students are "full members" (p. 260) of the academy. Interdependence is also an acknowledgement of what FG students bring to the academy and what they offer to the process of knowledge construction.

LIFELINES

Students also spoke about the extent to which their sense of place and belonging was sustained by their peers and their faculty or advisor networks, particularly when they hit moments of crisis or low points in their college journey. They described specific individuals, members of their cohort, MLC faculty, and advisors as lifelines who brought them through difficult moments through direct or indirect support. For example, Ruben described the role that his MLC peer group played in helping him stay on top of his academics while he was juggling school and employment:

> It was getting to the point freshman year when I just couldn't find a job, finances were thin, and I was trying to figure out the best way for me to continue going through college. I ended up taking a part-time job, but it was causing me to miss out on my study time. So Christina, Gregory,

> Nekisha, Tyrone, and Mona . . . everybody made sure I was OK—it wasn't like my hand was being held. But they would call and check up on me and make sure I was OK or to make sure that I finished part 1 of the reading. They did it in a very sneaky, surreptitious way where it was like, "Oh, I read this, what did you think about this?" to make sure that I knew the readings. They were always, always, always there.

Ruben reflected on how this support from his peer group and their caring but intrusive presence continued into his college career. He also acknowledged that he views the faculty members to be as integral to the community as his peers were, and he came to expect the MLC faculty to honestly appraise his progress in academia. He said,

> I think about the experience I had with working with the MLC instructors and how we knew what was going on with everybody the whole time. It was annoying at the beginning, but it became so, so instrumental to who we became. We still leaned on each other the entire college career. I will never forget those guys, and when I say those guys and say cohort, I am including the instructors as well. I think a significant factor of why I remained in school is because I had people around me to tell me the good, the bad, and the ugly no matter what. It came from the faculty and . . . and it came from my cohort at *all* times [Laughs]. Junior year there was a *lot* going on, and they always refocused me and were like, you know, "Ruben—Focus! You always say that to us, and you always pushed us, and now it is our time to push you and let us help you." Because I am stubborn like that, but they know me well enough to know that I needed help, and they were there for me that was a huge factor of me staying college . . . just having people tell me the honest to God truth about situations. They were always honest with me and I can appreciate that. Always. I have never been without resources, ever. I feel that I will never go through life without resources again because I always got my cohort.

Anna built on Ruben's comments and addressed how the ongoing support of her cohort through the many ups and downs of her college career helped her keep her eyes on the end goal—graduation. Some of Anna's MLC peers graduated two years before she did and began to define themselves in new ways. Anna saw them as her role models and evidence that FG students can make it through college to new opportunities:

> Oh, I definitely look to some of them individuals as for support. I mean, one of them is in graduate school, the other one is off working, they've graduated and they've taken their place in society. They've

> started building on their lives. It is really nice to look and see . . . like, this is going to be me in a few years, this is going to be me soon. It is really nice to see that because, especially like, I am a first generation college student, so I don't have many friends or family members that I see doing that right now. So it's like, that's nice. . . .

Anna talked about a particularly challenging time when she was placed on academic probation and was suspended because she had to withdraw from a course. She described how members of her learning community cohort and faculty advisor were a source of comfort and helped her strategize how to return to school. She recalled,

> A low for me was when I had to withdraw from some of my classes one semester and I was on [academic] probation—so that suspended me from college. And then I had to appeal and go through that whole process and that was very hard for me. And I remember my learning community faculty advisor was there for me through that, and that was nice to have somebody there for me. That was probably a low point. I knew that I was in a circumstance that I couldn't get out of, but I wanted to accomplish what I had started, you know? I had put a lot of effort into school and being here—so I wanted to just finish. I honestly believe that if it probably wasn't for my advisor, I don't know who would have helped me with the paper work, who I would gone to talk to. I didn't know where to begin; I didn't know if my circumstances were valid, I didn't know how I should explain them. I had no idea. It was hardest because my fate relied on this form that I filled out. I remember the day that I came and picked up my answer and I was so happy, I was crying, I . . . it was just great to get another chance.

Anna's story captures an all-too-familiar situation that many FG students experience. In some cases, the criteria for academic enrollment or financial aid status are not fully understood, and when there is a crisis, students might find themselves without a needed loan, the right number of credits for a grant, or access to a particular major. In the absence of a support network and staff or family who can clarify their options, students might take the wrong action or may simply decide that this event is confirmation of what they have always suspected—that they do not belong in college.

Davu talked about his strong relationship with his TRIO academic advisor, a relationship that began in the MLC and continued throughout his career, even after he transferred into his major and was assigned a new advisor in a different college. He said of his TRIO advisor, "Nolan is like one of the greatest people. . . . I think he could ever inspire anybody and he was our advisor. Anything you needed he could, he made you feel welcome to ever ask. I am a fourth year and I don't even know my new advisor in my major, but I know Nolan and I get more help from Nolan [Laughs], you know? I think I talk to them because, like, they made me feel welcome in the first year. It's like they make you feel that you could do whatever you want—always pushing me, motivating me."

These students' comments reflect on the value of having meaningful relationships with supportive peers and staff who understand their journey, their strengths, and their foibles. Above all, students felt understood and aware that they had found networks of support they could lean on without fear of judgment. This is of particular relevance to FG students, who often feel like they have no one to turn to when things go sour at school. Lauren shared this sentiment when she said, "Well, I mean, my family was like proud . . . but there was no real support." As such, FG students in the MLC understood the need to capitalize on the support of insiders who could help them plot a course through challenges along the way. Ruben captured this perfectly when he said, "To know that when you feel that way that someone is always there—I don't know how to explain it—it is invaluable. These people will tell you the good, the bad, and the ugly at all times and that's what you need [Laughs]. The MLC faculty knows that they won't make any more money by telling me good things—they have no stock in it except for us."

DIVERSE PEERS AS CATALYSTS FOR NEW WAYS OF KNOWING

Students also talked about how the diversity of their peer group in the context of race, class, language, national origin, and gender intersected with the multicultural curriculum to create a sense of belonging for them on campus. Research on learning communities at LaGuardia Community College, which serves a large population of diverse FG students, underscores that a "conscious creation of community in the midst of diversity provides a necessary support system" (Smith, McGregor, Matthews, & Gabelnick, 2004, p. 191), particularly for students who are strangers to academia. While the MLC students shared the experience of being FG college students, it was through their discussions about social inequities that they began to see the many other points of connection and difference between their experiences as people of color, as low-income students, and as outsiders to higher education. Zahara differentiated between the intragroup diversity that she experienced in high school and the diversity of race, culture, and ethnicity that she experienced in her learning community: "What sticks out the most was the community part of it—like, it felt good. I was with a group of people I had never been with and was that close to somebody who was Somali or somebody who was Asian. I am familiar with my culture, but I am not familiar with other people's. So I think the best part was getting to know everybody's background."

This combination of diverse peers and a diverse curriculum pushed students to ask difficult questions about themselves and one another. The process of engaging authentically in dialogue about identity and community allowed students to come to a deeper understanding of one another. Ruben recalled how he began to engage in the curriculum and with his peers:

> And each course made us question, What does this mean to you? And I've had to ask myself, As a Black man, Ruben what do you think about this? It would help bring down some people's stereotypes about race and gender and location. I remember Rita told us a story when she moved to Minnesota and went to the mall that was the first she'd ever seen a Black guy in her life and she was so scared. And I am sitting in class thinking, like, "What is her problem, like, what are you scared of?" But we were able to flush things out together and figure out the different images we have of each other—it wasn't a negative thing to do this.

Stories and first-person narratives presented in curricular materials or shared in the classroom invited students to explore worlds that they had never encountered. For students in the MLC, sharing stories about their own personal experiences created a sense of place in which they were understood. The diversity of this peer group and the role this heterogeneity played in enhancing students' connections with one another underscore the value of both diversity and a multicultural curriculum in allowing students to find a sense of place. Paul's comment illustrated the value of storytelling and the trust it engendered in the learning process (Jehangir, 2009):

> Just the fact that people were willing to hear your stories, I mean, it was different in a way. I learned to appreciate not everybody is comfortable sharing their experiences. Because we are not taught to share our experiences with people. I mean, when I first got over there, I am like, "Why do I have to tell you my business? This is my business." Then you realize that, that's who you are, if you can't acknowledge, who's going to acknowledge it? You know what I am saying? I think that was

> the difference. I mean, people see differences, not everybody is used to differences, but I learned to appreciate it, because I knew something about people. Like, all the girls in the class and the guys in the class, from Dominik to Ananda to Sasha to Demitri to Jenny—I mean, we all knew something about each other . . . where we were from. Anyway, when we first started those subjects, people didn't want to talk about it because people were aware of their surroundings, who was looking over their shoulder and stuff like that. But after you have a relationship with somebody and they are ready to know you, they know that the things that you are saying . . . it's different. Yeah, we are from different backgrounds, we are always naïve to other people's culture. If you make a mistake and a stupid comment, someone will correct you.

Paul acknowledged that despite his early resistance, he recognized the value in coming to know and understand his peers. He spoke about how many educational spaces do not prepare people to share themselves deeply and expressed how the learning community encouraged him to ask questions, not only of others but also of himself. To come to this place of authentic engagement required cultivating a sense of trust and also a willingness to take some risks in the classroom—this was true for both the students and the teachers in the MLC and is explored in the following section.

Building Trust and Taking Risks

Many students felt a home in the MLC and this comfort facilitated a willingness to share ideas, take artistic risks, and challenge each other. It was not easy to get to this point as a group, and it required that faculty be as committed to being vulnerable and flexible as was expected from the students. Anna spoke to the role that faculty played in modeling and supporting self-reflection and vulnerability as part of learning: "I think because there was an emphasis put on the professors getting to know us, and that kind of empowered us to come out more and to display our characteristics more and to know what makes us mad, what makes us sad; we were allowed to be ourselves. That helped us, we were allowed to be us and to find out what worked for us and what didn't work for us. And I learned from it, definitely learned from it."

Anna's comments highlight the cognitive, social, and emotional dissonance that students experienced as they studied issues of racism, classism, and other inequities from theoretical, historical, and artistic perspectives. As students grappled with these topics, they also considered them in the context of their own lives and identities, which resulted in moments of anger, sadness, and frustration. Rather than subdue these emotions, the instructors in the MLC encouraged students to express these feelings verbally, in writing, and through various media of artistic expression. Having the room to express one's self helps to create validating spaces where students can engage in their multiple identities and make sense of their strengths and weaknesses as learners.

Building on Anna's reflection, Zahara recalled that her comfort and trust of peers and instructors allowed her to take more risks in sharing her ideas without the fear that she would be wrong. She noted that her instructor's expectations played a critical role in her confidence and assured her that her insights were valued: "I was able to just kind of be myself in those classes. And able to talk. That was one of the first classes that I actually was able to openly express my . . . how I felt, without worry, because sometimes I find myself not knowing how to say it or what to say. But in the MLC, for some reason, it just kind of flowed because I felt that the teachers were confident in what I was going to say."

Finally, students commented on how their sense of belonging and comfort extended beyond the classroom into individual interactions with instructors. For example, Davu mentioned how working closely with his MLC composition instructor influenced his confidence in the writing process. Central to these one-on-one interactions was not so much being told how or what to write but rather the process of dialoging with an instructor who gave merit to his core ideas and helped him develop and own his writing: "Like, you could be confused to a point where you are stuck, you can't go nowhere in your paper, and he just sits there and asks you the ideas that you have and the issues that we were talking about in class and you literally, like, come out with a topic for yourself and he shows you how you could approach it in different ways, but at the end of the day, you have some idea of what you can do and then I know most people thought, he's very, very helpful."

Clearly, individual interactions with teachers occur in many non-learning-community classes. Yet what many students highlighted was that these personalized interactions were an integral part of the values and strategies of the MLC. Building relationships with each other was integral to the philosophies that the instructors and the students shared in common and brought to the process of learning community. The strength in the learning community design is that individualized interaction is not an isolated or chance occurrence but an intentional part of learning together. Three to four years after their participation in the MLC, participants recall both their isolation when they first came to the university and the ways that the MLC helped challenge that isolation to create a sense of belonging on campus. There were many facets to this process of finding a sense of place. The learning community structure itself allowed more frequent and regular interactions with faculty and peers, but it was the intersection of this structure with the multicultural curriculum, critical pedagogy, and intentional partnerships between academic and student affairs that created a web of support for students.

Students were not simply invited into a community, they were asked to *build* the community—to make it their own, share of themselves, and engage in the care of one another. Despite the heterogeneity of their race, ethnic background, religious affiliation, gender, and other differences, students converged in their common experience of being the first in their families to attend college. And though they often had divergent views on topics of discussion, they were able to see the points of connection in their educational journeys and aspirations.

SEARCHING FOR PLACE ANEW

As previously indicated, the student voices illustrate ways in which the MLC experience helped many FG students gain a sense of belonging at a large research university. For many students, the experience of starting their college career with a critical mass of FG students created a core peer group that encouraged and understood their journey through college.

When interviewed after their participation in the MLC, several students spoke about their sense of community and belonging in the classroom as a past occurrence, something that they missed and unsuccessfully sought to find in other arenas of the university. Although many students maintained relationships with MLC faculty and staff beyond the first year of college, some found it harder to do this with peers, especially when the organizing structure of the learning community had dissipated and the demands of work, school, and life away from campus started to grow. Zahara described talking to a few MLC friends on the phone after her first year, but she said, "We don't keep in contact, we're all busy, we are all separated to our different colleges. . . . Nobody really talks much anymore."

Other students, however, successfully sought out campus organizations that validated their identities as students of color or immigrants, and they also sought out new peer groups and staff to create safe spaces. For example, several male students joined a historically Black fraternity that provided a supportive and empowered peer social network. Other students became involved with the Black

Student Union, La Raza, Hmong, and Ethiopian student groups. These groups were more homogenous than the MLC, but they were the most clearly identifiable campus sources of both support and activism for our students.

While some students successfully translated their connections from the MLC to new relationships with key faculty in their majors, many struggled to form meaningful connections with instructors on campus or found that faculty did little to encourage relationships with students. Paul said, "Right now, I have started building new relationships with other professors, but I tell you what, I haven't been doing so well in school. It's hard to focus on school when you have all these family issues running around. But this year, I have tried to have better relationships with my professors."

Some students, such as Davu, were more successful in making connections, particularly when they interacted with faculty who were interested in engaging students both intellectually and relationally. Davu spoke about his relationship with a faculty member in Afro American studies. He said,

> In this university, one of the people that most motivated me would have to be my Afro Studies teacher. She's like a grandmother material, she was basically, like, one of our MLC teachers, except an older version and she's White. [Laughs] And in her classroom Introduction to Africa, she let me express my opinion. I can express whatever and she allowed it. You know, there are these professors that don't want their curriculum to go a different direction or be offending some people. But her and our MLC profs just allow people to do it. I think that's one important thing—you find *yourself* when people support you.

Davu's comment speaks to the way in which a sense of belonging served as a valuable springboard for students to engage in a deeper understanding of themselves and to ask the questions about their identity, their relationships with others, and their learning styles. Although the literal sense of belonging to a particular peer group dissipated for some students after the first year of college, what they took from the experience was the realization that such learning spaces and relationships can exist. In the retrospective interviews, students expressed how they continued to perceive the MLC as a validating space that allowed them to consider core beliefs, motivations, and interests that would carry them through their academic and life journeys.

The next chapter builds on this process and focuses on the theme of identity. While the concepts of belonging and identity are presented in two separate chapters, they are not mutually exclusive. As students came to a deeper understanding of themselves, having a safe place in which to expose their private and developing concepts of self was central to their identity development." Frank Lloyd Wright's quote at the beginning of the chapter alludes to this idea of interconnectivity—just as the "house is not on the hill, but *of* the hill," so too is belonging important to and part of finding and claiming the self.

CHAPTER 7

CLAIMING SELF

IDENTITY AND ACADEMIC VOICE

Numerous scholars have drawn upon a range of theoretical models (from psychosocial to cognitive to typological) to explore how college students make meaning of their lives and the world around them. In this chapter, I use the framework of self-authorship to discuss the development of low-income (LI), first-generation (FG) students' social, academic, and relational identities in college. In the student interviews, two themes salient to identity development emerged: *claiming self* and *academic voice*. In this book, critical and multicultural pedagogy were the guiding frameworks shaping the Multicultural Learning Community (MLC) and the *etic* (Stake, 1995) sentiments of the researchers and teachers involved in this project. As such, the interview analysis conducted by the MLC staff resided in a constructive paradigm and considered identity from a social-constructionism perspective (Weber, 1998). This approach considers identity as "socially, historically, politically, and culturally constructed at both the institutional and individual levels" (Abes, Jones, & McEwen, 2007, p. 2; Omni & Winant, 1994) and recognizes a fluidity to these social identities that shift with the context and relationships.

Claiming self refers to the ways in which students described how the MLC experience brought them to a "new awareness, development, and engagement with their multiple selves. While in many cases, race was the most overt identifier for students, many began to unpack the ways in which their complex identities in context of class, gender, language orientation, immigrant status, and family roles intersected with each other" (Jehangir, 2010b). In addition to intersections of social identity, students also saw overlap between their academic and social worlds. Thus *academic voice* pertains to the various processes by which students began to see themselves as learners and contributors to knowledge construction. Engaging in dialogue with peers and faculty, reflecting on provocative classroom moments, and working with curricula that connected with their lived experience impacted how they viewed themselves as learners and served as catalysts for their awareness of their academic identities. Therefore, claiming self and academic voice were interdependent, and students spoke to identity development with regard to not only their sociocultural identity but also how they perceived themselves as learners.

Given the multiple identities that FG students carry, this chapter addresses ways in which the MLC served as a springboard for students to make meaning and to find intersections between their many selves. Drawing upon the work of Kegan (1994) and Baxter Magolda (2001, 2007, 2008), these themes in the narratives of FG students are conceptualized through the dimensions of self-authorship. While explorations of self-authorship often include a discussion of four stages within each dimension—external formulas, crossroads, becoming an author, and internal foundations—for this work, I focus on the three broader dimensions in each student's narrative: intrapersonal, interpersonal, and cognitive development (Baxter Magolda, 1992, 2001). Attention to these dimensions in the interview data allows the student voices to demonstrate the intersections and the often nonlinear development among these three areas.

SELF-AUTHORSHIP AND FIRST-GENERATION STUDENTS

The concept of self-authorship or holistic development was introduced by Kegan (1994) to illustrate the intersecting nature of an individual's cognitive, intrapersonal, and interpersonal development. Building on this concept and with specific attention to the development of college students and young adults, Baxter Magolda (2001, 2007) defines self-authorship as "an ability to construct knowledge in a contextual world, an ability to construct an internal identity separate from external influences, and an ability to engage in relationships without losing one's internal identity" (Baxter Magolda, 1999, p. 12). The intrapersonal dimension pertains to how one views one's self in relation to the world; the interpersonal domain pertains to how one engages in and views social relationships with others; and the cognitive dimension addresses one's epistemological view of the world (Baxter Magolda, 2008).

Many students come to college with dualistic ways of knowing, seeking affirmation or definitive answers to questions from sources external to themselves, such as their parents, teachers, or religion (Perry, 1999; Pizzolato, 2003). The goal of a college education is to create opportunities for students to begin thinking in contextual, relativistic, and intercultural ways as they prepare for citizenship in a global world. How, then, do students begin to ask themselves the significant questions: Who am I? What types of relationships do I want to engage in with others? How do I know? One critical component in developing self-authorship is to experience disequilibrium as a result of "provocative experiences" that push students outside of their comfort zones to consider new ways of thinking, being, and engaging with others. In the context of this study, ***disequilibrium*** refers to the process in which students grappled with new ideas, new peers, and the larger academic context. This process involved conflict, both internal and external, as students tried to digest, absorb, and reflect on issues of race, class, gender, and inequality with regard to their own identity, the curriculum, and the voices of their peers (Jehangir, 2008, 2009a).

While studies of predominantly White, traditional college students suggest that self-authorship does not develop until well into adulthood (Baxter Magolda, 2001), there is evidence to suggest that historically marginalized students in the academy are better positioned to develop self-authorship prior to their early 20s. Pizzolato's (2003, 2004) work with high-risk college students, many of whom were FG students of color, demonstrated that the extent to which students were able to develop strong self-authorship pertained to how effectively they coped with the challenges of college. In many cases, students were able to draw upon their prior experience with ambiguity and were willing to process the disequilibrium created by uncomfortable experiences to cultivate a clearer sense of personhood. Similarly, Torres and Hernandez's (2007) studies of ethnic identity and self-authorship found that some Latino and Latina students were able to build strong internal foundations that helped them sustain their internal identities across the diverse contexts of their home and school lives.

For FG students who have a limited understanding of higher education, the entire college experience may be viewed as a series of provocative experiences that evoke questions about their place or value in the academy. Already engaged in a dualistic experience between their home and school lives, FG students need safe spaces in which to make meaning of these provocative experiences in the context of their multiple selves. If disequilibrium results in forced acculturation to college culture, then it is presumed that students are relying on external messages to shape themselves. However, when students have intentionally designed educational opportunities to consider their multiple identities as intersecting parts of a complex whole, "the self has added a third element which is greater than the sum of its severed parts. That element is a new consciousness" (Anzaldúa, 1999, pp. 101–102). As the many narratives of FG students suggest, they are well equipped with life experiences that raise questions and create dilemmas about their identities, roles, and purpose in higher education. They are familiar with switching in and out of different personas to meet the needs of different contexts—parent, caretaker, student, and cultural broker, to name a few. Therefore, what these students need are learning environments that draw upon these strengths and create opportunities for them to reflect upon their challenges so that they can incorporate their social and cultural capital into the academy.

While no one class or educational experience can be credited for developing the self-authorship of any student, critical learning experiences can begin the process by which students begin to ask themselves major questions about their values, beliefs, and ways of knowing. In the following section, the narratives of the seven featured MLC participants highlight the themes of claiming self and cultivating academic voice in the context of intrapersonal, interpersonal, and cognitive self-authorship development.

Student voices are presented in three sections to reflect the three dimensions of self-authorship: intrapersonal, interpersonal, and cognitive. A final section discusses integration across these areas and addresses the future aspirations of these students, particularly because these aspirations demonstrate ways in which the students crystallized their core values. It is important to note that disequilibrium was central to each dimension as the students struggled both internally and with one another to make meaning of new ideas, new relationships, and new understandings of the self.

INTRAPERSONAL DEVELOPMENT

As students reflected back on their time in the MLC, they focused on the ways in which the experience had raised major questions for them as individuals. The central question that students grappled with was "Who am I?" Students spoke about the way the multicultural curriculum with attention to multiple identities, coupled with a pedagogy that demanded reflection, put them in a position to consider, examine, or reexamine their sense of self.

Awareness of Stereotypes in Context of Self

For many students, examining their racial, ethnic, and cultural identity was an early part of understanding self. Students expressed an understanding of how stereotypes had shaped their perceptions of self, as well as how others perceived them. They were in the process of making sense of this awareness, and the journey was, at times, exhilarating while at other times emotionally draining. Lauren described how the curriculum and classroom dialogue prompted her to revisit her own history and family ties and try to better understand her own personal strengths and weaknesses in light of this reflection:

> As a community, we really attacked a lot of social issues, and [the MLC curriculum] made me aware of where we are racially, still, in this era and in this time. It made me *see* myself and look back into my past. And that right there just kind of captured my attention, because I never really identified with my own culture. So it kind of inspired me to explore the history. It made me embrace the cultures that I was born into. So it made me look at my family a little bit different, as far as their personality, the way they celebrate. And I became more open minded about just freedom of expression, period.

Lauren's comments addresses how culturally validating learning can provide both enculturation and acculturation to the academy (Ortiz, 2000). This process of self-examination was not without challenges; in fact, for many students, it brought back memories of earlier experiences that they had struggled with but never resolved. Moreover, it gave the students opportunities to do the hard work of understanding their emotions and thought processes surrounding these critical issues.

Davu recalled his anger as he came to consider the experience of immigrants and people of color, particularly in educational contexts. He recognized that anger was deeply embedded in all aspects of his learning. While it was challenging to go through this during his first year in college, when he viewed it retrospectively, he appreciated having an outlet for this anger in the MLC: "I was very opinionated; I always raised my hand; I always had something to say. It might have come out in an angry manner. It was very nice to actually see it come out in my papers. I know one of the girls used to call me Malcolm X, but you know, I was very opinionated. I liked what I was writing, and my anger was coming out in my writing. At the same point, I learned a lot about myself."

Villaverde (2007) suggests that engaging in a multicultural curriculum creates stages of critical awareness, one of which is anger or "a general frustration with frequency of discrimination, injustice and disregard for human life" (p. 208).

Davu also discussed being able to process this anger and express it in writing, discussion, and reflection. While he acknowledged his frustration with addressing stereotypes, he also found a way to embrace his anger and to claim it as valid and critical to defining himself. Villaverde (2007) notes that *contemplation*, or the act of reflection, is critical to working through this anger and considering one's options for future praxis. Davu spoke of how this anger and dissonance were necessary ingredients that helped him understand what he valued and how he wanted to represent himself:

> I think that, for me, it was just that the multicultural issues directed me toward my true qualities. It showed me the issues, and I finally found what I was interested in. What it did is, it literally showed me, showed the path, because I figured out who I was as a person, as a student, and you know, as a citizen. Just literally I figured out what I wanted to do my first year or so. It paved a way for my major, what I wanted to do, the issues I cared about, or hold deeply, and just it brought out my inner personality—so it helped me figure out myself.

Jarod also addressed how his learning experience pushed him to think about his own identity. He talked about reevaluating how he viewed himself when he recognized and challenged stereotypes that he felt constrained by: "It made me think a lot about who I am. And why I'm here on the planet, why I'm in school. You know all of these different things. I think being so much older than everyone else, that kind of had an impact. I kind of fit into a lot of stereotypes . . . the White heterosexual male, the kid from the single mother, and now they kind of shape who I am. I knew it before, I guess. But I was less conscious about it. I mean, now these things are a part of me, part of who I am."

Parks (2000) suggests that "if ongoing development is partially prompted by conscious conflict, then big questions play an important role" (p. 137) in addressing the conflict. Both Davu and Jarod's comments illustrate that these big questions, which emerged in light of the MLC curriculum, prompted self-reflection. These questions were rooted in the process of students' efforts to make sense of themselves and ultimately in the question "Who do I want to become?" (p. 137). While the students may not have arrived at all the answers immediately, they were beginning to navigate these complex questions.

Applying New Learning to Lived Experience

Thinking about classroom discussions that juxtaposed the complex issues of identity with regard to skin color, power, privilege, and gender, Zahara recalled being challenged to peel back the layers of her many identities. For Zahara and others, the question, who do I want to become? involved reconciling conflicts between how others perceived them and how they sought to define themselves. Reflecting on the long history of biracial bloodlines in her family, she addressed the struggle to claim both her Blackness and her identity as a student within and outside of her own community. Zahara said,

> I know I had a point there where I was [struggling] with who I was, period. Like I said, it made me understand myself more, like it's OK with being who I am. I was coming to grips with the fact that I'm Black, but I don't look Black . . . that whole issue of being light-skin African American. Like, even being Black, you still have prejudice within the Black community, so not only do I fight, not only do I have issues as far as maybe being a part of or fitting in, issues with what my body is supposed to look like, body image and things like that; I also have issues within my own community. As far as I am not Black enough, you know, maybe I don't talk like them, maybe now I am educated, now you know I am better than them. I had to get to know myself and being OK with who I am.

Zahara's comments demonstrate how the intersection of multiple selves creates internal conflicts, particularly in relation to how one filters meaning about each aspect of identity in various contexts. (Abes et al., 2007). Her struggle to claim her identity was shaped by her perception of exclusion from her community as a result of not only her skin color but also her education. For her, "not being Black enough" extended from her physical appearance to her educational choices, which seemed to distance her further from her community. She expressed the challenges of straddling two disparate worlds:

> Within the Black community, everybody isn't just Black. You're high yellow, you're light skinned, you're mixed, you're brown skinned, caramel skinned, you know, you're Black. You know what I mean? That's one of the questions that it raised for me when I was in that learning community, being who I was, because I had issues with that. Like fitting in with one side of town and then being with this other side of town and having to conform to each group when I am with them. Because when I am with people who are more scholarly and educated and when I am with my friends, or we are in the city, or I am in my community, I have to be another person.

Ruben's experience as a person of color in graduate school is similar to Zahara's in that both were trying to voice the complexities of their identities and resist defining themselves in terms of how others might seek to define them. Both students were also negotiating with different contextual influences, from communal or family norms to stereotypes and sociopolitical conditions, that impacted their views of the self. Ruben reflected on his experience in his master's program:

> As I sit here in graduate school, it is hard being an African American. As a Black male out here and in a graduate program, so often it's not just about being you. Now in graduate school from eight to eight, I am with my cohort; we're doing practicum, and then from here, I go straight to school. Anytime that I go out, we're doing happy hour with the cohort, and I feel like Ruben is being suffocated, like I am not allowed to be me because I always have to talk "scholarly." I feel like I had two identities . . . That's exactly how I feel. I am 23, everybody else is 26 or 27, and I like to have fun. I am a goofy child; I don't want to be too immature. But there is a reason why you are out there; maybe they need to hear that goofy voice. And that is something that I will always take with me in everything I do—just don't tell me that I should apologize for being me, no matter what it is.

Unlike Zahara, who said she is perceived as not being Black enough, Ruben struggled with how to be an authentic scholar without stifling his identity as a Black man. Both are acutely aware of walking the fine line between who they are and how they must present themselves in both their home and school lives. This process of deconstructing personal experience to find voice and engage in advocacy for self was a critical turning point for each of them in the process of moving through their dissonance and toward a more holistic concept of self. Ruben talked about coming to this junction and actively using self-reflection to inform his choices:

> I made sure that even in graduate school—I mean, that is the big one—wherever I am going and whatever it is that I do, that I am going to be doing things that will allow me to be me. Because I know how important multiculturalism is and a community is. I understand that each person is multicultural; you know, everybody has their own experiences. But if I am in a situation that it going to suppress me, I know it is not going to be good for me. I know now when I am interviewing or when I am looking for an apartment, I have to be at a place that is going to allow me to have that outlet, be able to speak up when I need to speak, you know?

Ruben's reflection serves as a springboard for the next section, where students describe how they worked to find congruence with seemingly incongruent parts of their many selves.

Choices and Actions: Finding Congruence between Multiple Identities

For many of these students, the MLC experience became a catalyst and conduit by which they worked through dilemmas about identity. While many students began by naming and challenging racial or ethnic stereotypes they had experienced, they also began to see how many aspects of their identity intersected with each other.

Lauren said that her frustrations with the new ideas she encountered in the MLC curricula and the conflicts that arose between this academic work and her own view of herself pushed her to ask questions about what she truly valued:

> I think, for me, it really helped me get in touch with myself as an individual. Even though we were on a collective level, it helped you look at yourself, consider what you think about life, where you stand in this world, and things like that . . . but also how you choose to express yourself. So it was kind of almost like a personal awareness thing for me. And so it really made me feel like I can take risks. I went from a major in business law and having a love for numbers to feeling more confident in taking more risks in the artistic side, the more free side. So that's why I did end up choosing to go to Europe. The change in my thinking, it just made me feel so good and so liberated. Rather than "Well, I'm just going to go get this kind of a degree so I can make all this kind of money." No, I wanted to be in touch with what I'm doing, and that is kind of what happened for me.

Lauren found herself moving from a life position that had been predicated on external expectations to allowing herself to consider other ways of thinking, being, and seeing herself. She spoke about her early marriage and life decisions as actions that limited how she thought about herself. While she had previously avoided situations or experiences that might challenge her worldview, in the MLC she was immersed in dilemmas about identity and community. Lauren described how, after initial resistance, she not only became more interested in exploring a multiracial identity; she also began to open herself up to interests that surprised her:

> Like I mentioned, I felt liberated completely. And that's just my personal experience probably because of my age and my circumstances at home. And when I returned back to college and was able to identify with myself . . . almost finding myself at a very late time in my life and in such a comfortable way where I felt lucky. So I realized that . . . I married my husband out of high school, well, not out of high school, but my high

> school sweetheart. So my life was taking a turn to focus more on Lauren, what she is, and what she wants to do in life, and skills and talents. I began to see myself as an individual and not as part of someone else. And so the liberation that happened with me gave me the courage to step out and, like I said, go to Europe and study, even though [my husband and I] disagreed about it. Which led eventually to my divorce, which I saw coming. When I came back to college, it was like I had to make a choice for myself or to live for, like, social status or something.

While Lauren openly acknowledged the discomfort that these new realizations brought to her life, she also credited the MLC experience as a catalyst in helping her gain a more reflective understanding of herself and who she hoped to become. In breaking down her personal experiences in the context of academic material, reflection, and discussion, Lauren was able to move away from limitations that she had placed on herself and could make difficult and risky decisions about changing the trajectory of her life. As she struggled with new ways of thinking about herself and her world, she came to see herself as a complex, multifaceted individual—a multiracial woman, an artist, and an advocate.

As these students' voices demonstrate, responding to the question "Who am I?" does not exist in a vacuum, and no one is one ever completely finished answering it. Rather, claiming self means going through the difficult process of asking the question and muddling through the many contextual, historical, and social forces that may apply while finding the answers. When students ask these questions, they must develop a compass by which they can navigate college and life with a confidence and awareness allowing them to make choices that reflect congruence between their multiple identities. While this is critical to all students, it is especially important to FG students, who often find themselves labeled or identified in ways that undermine their complexity. Paul captured the depth and importance of "recognizing the fluidity and interrelatedness of his many selves" (Jehangir, 2010a):

> Well, I guess my idea of identity . . . it goes back to what your definition of identity is. I guess I would be more leaning more toward perception and ideas about certain culture or just a certain reality, whereas before the MLC, my reality was different as compared to my experience. Having experienced that class, my reality became something else. I identify myself as a lot of things based on what I am involved in, based on what I have going on in my life, based on where I am at, so my identity is very diversified. Like, I can be a Liberian today, and African American tomorrow, I can be a Black person, I can be a Christian, I can be a meditating person, I can be a frat boy tomorrow, and I can be a soccer player, too . . . so it's like, so many words, I mean, just I can be a

> student tomorrow, my identities. I identify myself as a lot of things, but at the same time, too . . . I am just Paul, that's who I am—Paul.

Paul's quote captured the complexity of identity development in young adulthood and underscored the importance for FG students to be able to recognize and honor their many selves as they navigate the academy. Baxter Magolda (2004) argues that having the capacity to cultivate an integrated identity is characterized by "understanding one's own particular history, confidence, and capacity for autonomy and connection" (p. 6). This form of self-understanding is important for all students but is especially critical for FG students who often find their developing identities contested in both their home and school worlds. As such, curricular environments that facilitate asking questions about one's place in the world and that support the exploration of these questions can support FG students' development and capacity to navigate their future college years. The question "Who am I?" also gives rise to other questions about one's place in the world and in relation to others. The next section discusses these additional questions and contains student reflections on how aspects of their interpersonal development were facilitated in the MLC.

Interpersonal Development

At the same time that the students tried to make sense of their multiple identities, they also considered what types of relationships they wanted with others. Although all students in the MLC were FG college students, they were also a heterogeneous group representing different cultures, ethnicities, racial backgrounds, high school experiences, life roles, and so on. Engaging with one another in a learning community that examined these very differences from sociological, historical, and artistic frames resulted in many challenging moments for students. To make sense of this complexity, students could not simply hold onto their preconceived notions of themselves; they also had to ask, Who am I in relation to others? How do my relationships with others reflect my values and my motivations? As with intrapersonal development, students had to work through several waves of disequilibrium to arrive at a place where they could build meaningful relationships with others.

Widening the Circle: Creating Space to Learn New Points of View

In their interviews, many students shared what it was like to be immersed in deep and challenging dialogues with each other. They acknowledged that the ways they had previously organized their own, personal worldviews were being challenged not only by what their peers said or did but also by who they were. Davu talked about the value of having open dialogue, but he also conceded that he was sometimes surprised or shocked by the points of view of his peers: "These are things that are always going to be in this society, especially in America with the vast different cultures and language and religions. . . . It's always going to be there, but we never talk about it in a sense. It is always, like, oh, OK, it's always underneath the rug, but in the MLC classrooms, it came out. I mean, it was nice to actually see people expressing things; it was always people that were throwing a comment or something, and it would just literally take you aback. . . . What?"

Davu's comment speaks to the ways students learned to consider and respect the perspectives of others, even if they did not always agree with them. As the semester progressed, students recalled being empowered to use their own points of view as a communication tool during their discourse with faculty and peers. Zahara addressed the importance of having a space in which to engage in deep dialogue about issues of race, class, and gender. She said that the process of considering these issues from theoretical positions was always embedded with discussions about students' lived experiences and sense of community:

> So when we talked about White privilege in our class, it got a little hectic because there were only two White people in our class. And so I remember Jacob was kind of like, it's not my fault! I think that we all felt that we were learning something and he really felt it too. He was almost like teary like, I am sorry. We are looking at him like, it's OK. You know what I mean? We're having this big debate, and we're "But it's OK," because he was in our community, and so we felt, you know, we felt him. And we are arguing, everyone's in the class arguing. That was the best part of it—we all had our opinions. At the end of the day, we all reflected on what we talked about, and we were able to understand each other. So I think that was best part, the deeper we got into it.

Zahara's reflection on the difficult discussion of White privilege demonstrates how the themes of belonging and interpersonal development are closely linked. Students trusted one another enough to engage in difficult discussions, but there was also a sense of continuing to take care of all community members in light of these discussions. Ortiz (1995) suggests that informal peer interaction and discussion led by students in multicultural classes are effective means of learning for all students, but first-year students seem to benefit most from this model. This balance of voice in community illustrates the high impact of "enriching and supportive" learning environments (Kuh, Kinzie, Schuh, & Whitt, 2005) where students can be simultaneously challenged and supported by teachers, curriculum, and each other.

Community Conflicts: Safe Spaces for Engaged Disagreement

Zahara's aforementioned observations underscore the need for a sense of trust and community to allow students to engage in thoughtful disagreement. Zahara recalled the importance of having community ground rules and coming to know, understand, and respect her peers so that these discussions could be challenging without being damaging:

> But we set rules for us in the beginning—these were the guidelines for the community, and you respect what people say. In the community, we had assignments where we had to find out where were we from, what was our background, who was important to us in our family. And, you know, things like that, to share with each other, because, if you don't share, you have like no common ground. If nobody knows nothing about you, it's like, "What's the purpose of it?" You can't really talk if you don't, you know what I am trying to say? You can't understand somebody.

Building a sense of belonging and buy in as a community was essential to meaningful dialogue and exchange of ideas among students. This demonstrates the ways in which the dimensions of self-authorship are interrelated and overlapping. Paul recognized that to really begin understanding different perspectives was to experience frustration and dissonance and to commit to the process of making meaning from this conflict. Describing the process of making a mural with his peers, he said,

> Frustration, I mean, it's human differences; frustration always has to be part of life. We are human, we have emotion. When you have 30 people in your room, with all these emotions and passion in what they want to do, there is always going to be a frustration; not everyone is going to agree with each other. Now the fact that we were working on a project, that all of us had to contribute, everybody wanted to be represented on this project. Everybody wanted to be a part of it. It was a really, really good job. I don't regret the frustration, I don't regret the emotional tie, I don't regret the arguing. Because you can tell that everybody wanted to be represented on that mural, whether it be their painting, their pictures, their words, their quotes, their time, their voice. People just had to listen to each other. But at first, it wasn't

> easy, it wasn't easy—even to the last day, it wasn't easy. But when the mural was finally up there, it was like, Wow! We did it. So there was a lot of frustration, there was a lot of arguing, there was a lot of disagreement, but at the same, it was the disagreeing that helped make what the mural is. Because everything that happened, happened for a reason, even the things that went wrong. Actually, it is expected to happen; disagreements are supposed to happen. When you have a bunch of people together, they are going to argue.

Zahara and Paul both articulated how engagement with others, especially when this engagement prompts disagreement, is facilitated by having common ground with each other. For MLC students, this common ground was not only the "ground rules"; it was also created by their personal narratives, which had become interwoven into a larger connected narrative of community. As such, discussions of immigrant issues or race were no longer separate or abstract issues but were represented in the students' own life experiences and in their relationships with one another. Thus "students learn best about other cultures when they experience them directly" (Hornak & Ortiz, 2004, p. 91), and this engagement embedded in their lived experience allowed them to consider the question, Who do I want to be in my relationships with others?

Valuing and Seeking Out Relationships with Diverse Peers

Students also addressed the ways in which they began to value relationships with diverse individuals and to draw upon this support in order to thrive in their school environment. Anna talked about how she saw her own educational journey reflected in her peer group and found their support to be instrumental in helping her understand her own identity:

> Because there are other people that have probably been through the same thing that I have been through, the same issues that I am having. . . . I maybe even still have issues with claiming myself, like this is me, and this is who I am. I think that going through the MLC group helped me do that—just discover who I am as an individual and what I want to do. I just feel like I gained that from being around so many other people that had big goals and came from places like I came from, and look at them now, you know?

Like Anna, Zahara noted that the diversity of her peer group and their different attitudes and perspectives allowed her to see that she did not have to pigeonhole her identity. Rather, she could embrace many aspects of herself without having to constantly defend herself. In many cases, interpersonal interactions with diverse peers provided students with different models of being in the world and opened new possibilities for how and what they valued in relationships and for what they were beginning to value about themselves. Zahara said, "What I mean is because we are all different people from all different places, getting to know how this person felt, and how that person felt, the attitudes each person has about what's going on. I don't know, it kind of helped me to learn, know who I was, like, I am who I am and accepting myself for who I was and not having to maybe even justify who I am. You know what I mean—not having to explain."

Embedded in both Zahara and Anna's words is the way in which they perceived their community to be an extension of themselves. Their value of interdependence and connectedness was often in stark contrast to the more individualistic culture of higher education. Parks (2000) notes that "because the notion of independence is so strong in Western society and in the canons of adult psychology, if the need for family and community is strong it may appear to contradict the achievement of adulthood" (p. 91). For many FG students, this disconnect between the individuation of higher education and a more collective vision of the self can be especially challenging. As such, learning communities are curricular spaces that can help mediate and close these gaps, especially if the curriculum and pedagogy capitalizes on the strengths of learning in a community.

Reexamining Our Relationships to Others

Students also talked about ways in which the curriculum coupled with the pedagogy of engagement with diverse peers allowed them to build relationships and points of intersection between their lived experiences and academic lives (Jehangir, 2010c). For example, Ruben recalled finding common ground with an Asian female student who had been verbally assaulted and spit on while traveling to class on a city bus. Ruben described how the MLC curriculum and pedagogy created links between lived experience and the classroom in ways that allowed Rita to share her story:

> I remember 9/11 happened, and Rita came in to class, and I don't think she knew exactly what happened. She just knew that Americans got attacked, and I think they spit on her on the city bus and we had gone into this brown alert on campus—and me being a Black, it was

> just interesting because all of the sudden, me and Rita found ourselves in a situation of similarity because we were afforded the chance to get to know one another and talk about it. Had we not been in that learning community and a class in the liberal arts college, it wouldn't have been an issue because you got all of these kids there and have nothing to do with your personal life and academics but just about getting the knowledge and regurgitating it back.

Ruben's comment illustrated the ways in which he began to see connections between systems of discrimination and oppression across groups. Making these connections allowed students to break through prejudices or perceptions they might have held about different individuals and to develop "interdependent relationships with diverse others" (Baxter Magolda, 2007, p. 69). As students continued to consider points of connection and difference between themselves and others, they engaged in a process of "building and re-evaluating relationships in keeping with the multicultural learner perspective" (Torres & Hernandez, 2007). During the course of the MLC, students were challenged by diverse and often difficult multicultural issues. Discussions with peers around racism, sexism, homophobia, and inter- and intragroup prejudice pushed students to confront their own prejudices. The following three students describe how they came to reevaluate their prejudices and perceptions of the gay, lesbian, bisexual, and transgender (GLBT) community. When discussions in class intersected with her personal life, Zahara found herself trying to understand the position of gay individuals. She described how thinking, writing, and discussing homosexuality in the MLC helped her understand the experiences of her cousin and younger sister, both of whom had come out of the closet:

> We would all say that it changed us somehow, someway. It opened our eyes to a lot of things. I think that was my first multicultural class, and I have had a lot since then, but I think that was one where we really broke down being bisexual or homosexual. And I had a cousin who actually had just come out that semester, too, so I was able to talk about how my cousin came out to me . . . and so accepting it and having to accept it at my own family. I wrote about that in one of our papers; I think that helped me also understand my cousin, by writing it. And I was able to ask questions in class because it was kind of like taboo in my family. But it was just something that we just had to accept and understand, getting more details on, reading what people say and all the stereotypes and the discrimination and all that. My little sister who is 16 came out and said she is a lesbian. I had to understand.

Like Zahara, Ruben described how the classroom discussion and materials pushed him through the wall he had put up by turning away from his gay brother:

> My appreciation for different cultures definitely goes to a credit to the program because it was more or less, I had the army policy—you know, "Don't ask, don't tell—I don't want to see that." But it has totally, totally, totally changed me. I came into college—my brother is, actually, he's gay. And it was a hard time, and we were learning about [GLBT issues], and I had so much in me about the situation, and it poured out in all the classes and it allowed me write about it or to create something about it. It made me understand that there were so many different ways to express yourself. I don't know how to explain it; it made me get past what it was that I thought was bothering me, and I understood that I really missed my brother.

Unlike Ruben and Zahara, Paul did not have a personal connection to the GLBT curriculum. Instead, he talked about how his prejudices were shaped from an early age and how cultural taboos sanctioned this type of prejudice in his mind. For him, hearing the arguments of his peers and GLBT speakers allowed him to see new connections between their civil rights and the rights of other marginalized groups:

> I came into class, personally, with a very, very narrow-minded view in terms of GLBT. I come from a country where we don't play that kind of stuff over there, so like, my experience in that class just changed my entire perspective toward GLBT. Not that I hated them, but I just didn't want to have nothing to do with them because I felt that they are making the choice, they should take the consequences. But the more I read about it and the more we talked about the issue and just how bringing in speakers that were gay and of different sexual orientation. Just kind of switched my perspective in a lot of ways. I really learned from that class in a lot of ways and women's rights and stuff like that and just having to be lot more considerate of other's people's cultures. Because oftentimes we just try to generalize and try to put people in this little box and which is sort of uncalled for, but that's how society is; we try to put everybody in one box to make life easier for our self and not knowing that we are offending other people and disrespecting other people. You have to stand as an individual—so over the years, my experience in that class really shaped me in a lot of ways.

Paul's reflections demonstrate ways in which interpersonal engagement with diverse peers and curricula afforded him the opportunity to reevaluate how he views gays, women, and other groups. He and others addressed how these critical moments created a foundation from which to consider how one interacts with others. Students also began to consider how the beliefs one holds about others are closely connected with what one values as an individual. This intersection of intrapersonal and interpersonal development is augmented by ways in which the students asked the question "How do I know?" As Paul, Ruben and Zahara struggled to consider and understand the issues surrounding GLBT rights, they were also pushed to consider how they had learned to think about rights and equality. Therefore, the engagement with multicultural curriculum was connected to their cognitive development (King & Shuford 1996; Ortiz, 1995). The next section highlights ways in which these seven students explored how multiple ways of knowing shaped their learning paradigms and their sense of self.

COGNITIVE DEVELOPMENT

The cognitive dimension of self-authorship gauges how students know and process new information and how they engage in knowledge construction. As with the other dimensions of self-authorship, the cognitive frame intertwines with life experience and is stimulated by engaging with ideas, theoretical concepts, theories, or multiple modes of knowledge that challenge one's existing lens for viewing the world. For students in the MLC, this process involved asking the question "How do I know?"

Incorporating Multiple Ways of Knowing into the Learning Paradigm

Many students acknowledged that they often looked to external sources to make sense of their worlds. Working with a diverse group of peers and a curriculum that examined the social construction of race, class, gender, and other inequities challenged students to consider different ways of knowing. Lauren described her initial discomfort with ideas that caused her to question her existing beliefs:

> The one thing that did rise up for me as an issue was religious beliefs and personal morals and stuff like that. Because in the classroom, we were addressing issues that . . . a lot of times behind the actual issue is a religious belief or a racial or cultural belief. . . . So in the classroom, I think it kind of made me feel like, "Oh, well I'm a little afraid to be more open minded, because it might go against some of the things I've

been taught or whatever." So being challenged with that personal faith and being challenged to embrace other ways of seeing the world . . . as a whole rather than just my way.

Lauren's disequilibrium was echoed in the comments of other students, particularly as they came to recognize and question various inequities in society and to consider new theoretical and historical perspectives. Zahara's comment reflected a realization of how multiple sources of knowledge can bring new light to understanding complex issues:

The topics I learned educated me on different issues. So I am not oblivious. You are no longer in the dark about it. Now you know. You know what I mean, we talked about the American Indians, we talked about Adolf Hitler, we talked about all different races, and you know, people's experiences. So basically you won't leave the same that you came, because you won't be in the dark about everything; you have been opened up to anything. So I think that helped me be able to talk to people more. I think that was one of my beginning classes where I was starting to get more educated. Because you know you come into school and you know you are kind of naïve. So yeah, so I think that, yeah, the awareness of everything. That opened it . . . my eyes.

While each of these students struggled with dissonance prompted by new ideas and ways of seeing the world, working to accommodate these inconsistencies was a step toward a cognitive shift that recognizes how knowledge is dynamic and contextual.

Practicing Self-Reflection to Inform Choices

Davu also noted how critical it was to engage in constant reflection. Speaking of the reflective writing he completed in the MLC, he described how he moved from being irritated to gaining a better understanding of the process and progress of his thinking by reviewing and reevaluating his reflective writing:

It was just like every week you had to write a weekly response. And you'd sit there literally thinking about the class and what everybody talked about or whatever, and you would sit there and write an angry response. Just getting them back you would be like, "Man, I will read it tonight," and you'd set it aside. But you go home and then you try to read your words and the instructor's response and be like, "Dang, I said that . . . wow! I really said that!" I used to hate those reflections, but I kept them, I am not going lie, I kept them. And I still, to this day, years later, I have all of my responses for every week, and usually when I go through my boxes where I keep all my papers for my different classes and stuff, I would literally read it in and out. I was like dang, I was angry. By now, I know why I was angry.

In processing their cognitive dissonance, many students began to expand their own views, recognize multiple ways of knowing, and see the lived experience of others as valid sources of knowledge (Jehangir, 2010c). Lauren talked about the ways in which the MLC experience prompted her to explore her creative interests and to let go of her plan to major in business, which had always been a safe, but unsatisfying plan:

> It kind of opened my mind up to connecting different ways of expression through writing and visuals. And being able to be involved with the cultural aspect of it. Kind of put it all in perspective for me, and how I choose to, to express myself, the way I see politics, the way I see different things like that. And I think it shaped my decision to further my education in music, and things like that . . . theater, so it kind of sparked and cultivated things like that. It freed me to just embrace an intellectual side of myself, in a different way, in a different manner, rather than just studying a book.

Lauren's comments demonstrates how she began to critically reevaluate what constituted knowledge and incorporated different mediums of expressions into her frameworks as valid means of making sense of the world. Her ability to translate these new understandings into her own lived experience demonstrates how taking a multicultural perspective creates opportunities for a more cognitively complex thinking and meaning making (King & Shuford, 1996).

Recognizing Learning Environments That Are Conducive to One's Learning Style

Part of the process of meaning making in the cognitive dimension pertained to how students began to see and understand themselves as learners and co-constructers of knowledge. Students began to name and identify learning environments that were conducive to their learning styles.

Paul addressed how his intellectual engagement was enhanced in environments that built trust in the classroom. For Paul, this camaraderie extended beyond social connections to impact how he learned: "And you would be surprised how that sense of comfort zone actually motivates students to be involved their class; not only that, but being involved with their homework, being involved with doing their readings and stuff. Because the thing about it is that, if you show up to class and you aren't talking, the people are going to look like you didn't do your reading."

Paul's comments addressed how he began to create his own framework for learning, in which he valued both responsibility and opportunities for deep engagement with peers. Anna also noted that interdependent learning and involvement with her peers was more likely to engage her: "I think that participating in the MLC gave me a good sense of what it feels like to work in a group. Because we worked in a group often, I think that it has helped me be taking other people's opinions and thoughts under consideration more than I would have if I wouldn't have went through that. It was something I got used to. I probably would prefer to do that rather than sit and think to myself. I want to know what other people are thinking too, you know."

Lauren compared two very different learning environments: one computerized and the other classroom and discussion based. While she earned good grades in both experiences, she was able to articulate why one learning space was more beneficial to her than the other and in doing so reflected on how the interpersonal environment impacted her cognitive engagement:

> I had psychology of human development and that was computerized, as well as my algebra class was computerized. So it was all on the computer. And I did really well, but I think that . . . I keep mentioning the interaction with everybody—it just kind of creates some kind of energy between each other. 'Cause otherwise if you're sitting behind the computer and doing your homework and you're studying—you have no interaction with nobody but the screen and your own thoughts or whatever. So it's kind of like . . . I think the difference is it helps to have more human interaction. . . . I've discovered that through the MLC. Because I thought I was OK with the computer thing: "Leave me alone and let me study." Then after I had the experience, I can see the contrast between the two. And I'm like, "Yeah . . . I did good . . . but I could have done better."

As the students were gaining awareness of their academic voices, they also began to consider different modes of communication and to understand the nuances involved in communicating with different audiences while staying true to their intentions. Davu talked about this process of learning to adapt his voice to effectively communicate with a range of people: "I remember writing response letters to the newspapers, and I used to have my MLC instructor read my letters that I used to write. The [language] in the letters came out angrier, just as my papers came out angry. Our MLC teachers would say, you know, there's a better way of saying that kind of thing. I mean it makes sense. How I talk with my friends is not how I would talk professionally to other people in the business world."

Ruben commented on how these interactions with faculty allowed him to see writing as a medium for self-expression. He spoke of writing with a "purpose," not for the instructor or the grade, but for his own fulfillment: "Writing, my writing to me, it was a tremendous upgrade. I don't know how else to put it, after I left that program, I wrote for a purpose—everything that I have written has been for a purpose."

In learning how to compose texts for a range of situations, Davu and Ruben practiced how to translate their voice and ideas into new contexts. They also engaged in a process of appraising situations and being intentional and purposeful in their responses. For students like Davu and Ruben, the real asset of their writing experience is not completing a letter or an assignment but formulating a rationale about how this action connects with their voice and beliefs, which is a critical part of developing a knowledge framework as addressed in the next section.

Developing a Personal Knowledge Framework

Students addressed how their early experiences in the MLC set the stage for finding ways to express their voices and ideas as learners and teachers. Jarod talked about increased confidence in his self-identity but also about the merit and value of his ideas. how confident he became in his ideas and self-identity. He talked about practicing his academic voice in other classes and moving from the sidelines to become one of the most active participants:

> I mean, that confidence that I gained made it easier for me to say these things and write these things . . . and you know, even challenge professors about things . . . because of that MLC experience. In other classes, the things that I did in [the MLC] made it a lot easier to say, "No, this is not right." Like in high school, I was the one in the back of the class . . . you know . . . like falling asleep. And now it's like, there's a lot of classes where I'm the only person talking. I mean we'll have a discussion where [raises hand up in the air] "Fine, I'll answer the question." It's like I'm waiting for other people to speak. So if somebody had challenged me, I'm just not going to say, "Oh no, OK, I'll conform

> to your ideas because mine aren't good." So who I am is OK, so I'm going to try to figure out the world through those eyes and not just by trying to be another scholar.

Jarod's comment illustrates the way in which many MLC participants began to consistently engage with their learning environments in ways that reinforced meaning making. Students like Ruben talked about how they started to develop their own knowledge framework, allowing them to navigate and interpret knowledge and decisions in different contexts: "It made me take all my classes personally, in my writing, in the literature. I always tried to empathize with the setting or empathize with whatever was going on. Instead of regurgitating things back, I tried to make it meaningful. And I learned that all through the community." Both Ruben and Jarod articulate how they asserted their own purposes and goals in their learning and challenged the notion of knowledge as "banking" facts or ideas. Embedded in their narratives is an awareness that to "know" is not simply a cerebral act but one that involves engagement and empathy. This speaks to the ways in which the dimensions of self-authorship or claiming selfhood inform and shape each other. While individual students might lean toward expressing themselves in the language of one dimension over another, it is the integration and awareness of these three dimensions (intrapersonal, interpersonal, and cognitive) that illustrates one's journey toward developing a more defined sense of personhood.

INTEGRATED LEARNING: ASPIRATIONS AND ADVOCACY

Each of the three dimensions of self-authorship (intrapersonal, interpersonal, and cognitive) address the different components necessary for cultivating a deeper, more reflective, and engaged sense of personhood. Yet it is the way in which individuals navigate the complex intersections of thinking, feeling, reflecting, and being with others that shape self-authorship. Mezirow (1997) refers to this process of self-understanding as transformational learning and argues that it includes "critical reflection of one's assumptions, discourse to validate the critically reflective insight, and action" (1997, p. 60). This final section of this chapter highlights student voices that illustrate the connectivity between their intrapersonal, interpersonal, and cognitive realms.

The FG students in this study were not newcomers to the realities of racism, classism, sexism, or the confluence of these inequities in their own lives. They had, however, relatively few experiences that allowed them to deconstruct these experiences in conjunction with academic texts, artistic media, and narratives. Students who were interviewed three to four years after the MLC program expressed ways in which their first-year MLC experiences and other new learning experiences they had sought out since focused on social advocacy as a critical frame for the self. In an effort to put their new visions of the self into practice, they sought opportunities for multicultural engagement, often engaging in actions and choices that cultivated congruence among their multiple identities. These choices reflect the connectivity of intrapersonal, interpersonal, and cognitive ways of knowing in that the students addressed aspirations and actions that emerged from drawing upon their understanding of the self, their relationships with others, and their knowledge frameworks.

Several students shared specific examples of actions they have taken to create opportunities for others that embody their values. These values draw on how they see themselves, their roles in communities, and their understanding of longstanding inequities in educational access. Davu's advocacy focused on the youth in the Ethiopian immigrant community. His decision to serve as a mentor and guide to future generations of Ethiopian college students was a means of channeling his anger into meaningful action. He spoke with pride about being a "big brother" to young men in his community:

> I started this mentor program where I sent one college student on to church on Sundays. While the kids were at church, the church leaders gave me the classroom. There were about 20 or 30 kids and the kids asked a lot of questions, so college students respond to them, [such as] how to get to college, how you choose your major, how to stop with peer pressure with drinking and alcohol and partying and fun and having fun in school, working in school. So every two weeks I would send in one college student to that church classroom, I would actually be present too, and they would ask questions and we would answer. And these kids had all these opinions, and I took it down and I went to my board members and one of the events planned out that month was to bring them to the university for a day-long program. I took them all over campus and at the end of the day, I fed them and rented a bunch of bowling lanes, some pool tables and to have fun now, just like I am one of your brothers. It was very fulfilling. . . . I mentored like 27 kids, and seven of them graduated from high school and are going to college.

Most of the students interviewed expressed a keen understanding of the ways in which education is an avenue to upward mobility, but they also noted that this mobility has to be harnessed for the collective good of their larger communities. Zahara talked about her aspirations to build a community center and then incorporate health and social work outreach to communities and families. She differentiated between being viewed as a savior versus a resource:

> I don't want to be the saver, the person who saves everybody, but I want to be like a resource for people. I want to start an organization for adolescent youth as far as providing resources. My thought is to have this building with health services provided within the building. Kind of like a community center . . . Somewhere that provides maybe a three-room health care facility, like a clinic where there is inpatient-outpatient. Maybe a conference room where people can hold conferences. Things like that, also where we can do like family therapy sessions. Or you know, counsel rooms where people can come in and get consultation on marriage and even community people can utilize the facility. Ultimately that is what I want. You know, that's a big idea. So like a community similar to a learning community, that's kind of what I want my organization to be—like a community that learns from each other, but it somehow, it goes in a circle, so we all somehow, you know, help each other out.

Zahara's beliefs and values around advocacy and action draw a sharp line between support that is charitable and support that is rooted in the idea that each member of the community has something to offer to the collective. The inclusive, community-based aspects of her plan draw from her own intrapersonal struggle for inclusion in college and in her home life Her ability to work through her own dissonance and to see knowledge as dynamic and flexible also played a role in how she expressed her aspirations for advocacy.

Paul's journey through academia found him seeking spaces that allowed him to be "part of something." Engaged in organizations ranging from the Black Student Union, to organizing the annual Spring Jam event, to ROTC, Paul practiced finding places, people, and ideas that cultivated congruence among his multiple identities. Double majoring in sociology and youth studies, he worked with community youth organizations and found that like him, many young adults were seeking avenues for voice and creativity. Paul spoke about his volunteer work in youth programs: "I mean, young people want to have some kind of voice of what is going on in their life. People want to have some kind of responsibility, just giving them that responsibility, saying, You know what? Run this program. Oh, you have some idea about a summer camp? OK, come out with anything, anything they want to come out with, whether it is basketball camp, whether it's a dance program—these kids just come out with a program, design it for a whole month, and then run it."

His words echo the ways in which his own self-authorship developed as he broke from socially constructed formulas for viewing himself and others and came to embrace his multiple selves. In his work with youth, he recognized that they, too, are at varying stages of claiming themselves. He drew on his collective experience to help them facilitate their transitions.

The remaining two students, Anna and Lauren, shared hopes and plans for serving their communities with particular attention to the intersections of race, class, poverty, and mobility. Anna said,

> I have always fit in with a diverse crowd and been in groups with diverse multicultural people. I want to keep that going in some way; I want to stay connected, I want to keep doing things to help people that come from different backgrounds and that are disadvantaged. So I think that the MLC just kept me on track to doing that, and being a person of color as well has done that. I just want to, you know, give back to the community. My long-term goal is to start a company working with pregnant teenagers—pregnant teen women—and giving them life skills and providing them support. Help young women see that having a baby isn't the end of their lives and they don't get to go to college anymore. I want to help them see that they still can make it and be successful and have a bright future for themselves and their child.

Anna's comments reflected how her process of meaning making is rooted in an internal value and understanding of the challenges that young "disadvantaged" women face in getting to or completing college. Her life plans have been motivated by this process, coupled with her academic studies in family social science and her own commitment to seek out opportunities for multicultural engagement with others.

Lauren is the one student in the study who had left college at the time of the interview. One of the oldest students in the MLC cohort, she acknowledged experiencing a great deal of dissonance, especially intrapersonally and cognitively. Her journey through the MLC resulted in actions that changed the trajectory of her life. Although she did not continue in school, her decisions to study abroad, get divorced, move to a new city, and reinvent her life are evidence that she developed a strong internal compass for herself. At the time of the interview, she expressed interest in returning to school: "Currently, I wanted to come back to the university and do an independent study . . . individualized study degree. My interest is in transitional housing and homelessness. I'm in . . . my loans [are] in default, so I can't go back to school right now. So I'm trying to come up with a plan." While her financial situation limited her options to return to school, Lauren did start a nonprofit organization that provided transitional housing for the homeless.

Although many students in the MLC came to a deeper understanding of themselves and their place, motivation, and passions, some were still very much in the process of understanding what to do with this newfound knowledge of the self and the world around them. Jarod expressed a sense of disillusionment or letdown as he approached graduation from college. After returning to college from his study abroad experience in Italy, he questioned the point of acquiring the remaining credits needed to secure his degree. He said, "And then coming back I was really . . . almost like confused as to why I was here. It was like . . . this is just more book knowledge. This is more reading, and it doesn't have any real . . . I get it. I mean I don't want to take anything away from academia in general. I just kind of feel like I got to a point where I got all I could out of school. Whereas, 'All right I get it, I know what I'm doing here.' I know how to take the tools that I've learned in academia and apply them to my life."

Unlike other students who were able to find conduits for their passion for multicultural issues or community, Jarod found himself still searching for avenues of meaningful engagement around the issues he was passionate about. He was struggling to find ways to allow his cognitive self to intersect with other dimensions in his life. His comments are a good precursor to the final chapter in this book, which addresses students' critiques of both the MLC and higher education as a whole.

CHAPTER 8

ON CRITIQUES AND POSSIBILITIES

The academy is not paradise. But learning is a place where paradise can be created. The classroom with all its limitations remains a location of possibility.

—*bell hooks,* Teaching to Transgress: Education as the Practice of Freedom

The narratives of low-income (LI), first-generation (FG) students in this book reaffirm that college is a far cry from paradise for many students. However, these same narratives also include experiences and spaces that, though imperfect, provided an oasis of belonging and meaning making in the educational journey for many nontraditional students. These students' experiences and aspirations, from their first year and beyond, underscore the idea that the purpose of multicultural and critical pedagogy is to challenge educational inequality by creating possibilities and opportunities for all students. The classroom, and indeed, U.S. higher education "remains a location of possibility" to create democratic multicultural communities and challenge social inequities; it is *how* we, as educators, will nurture this possibility that is still a question.

In their book *Crossing the Finish Line*, Bowen, Chingos, and McPherson (2009) focus on the role of public and state four-year colleges in addressing educational disparities in the United States. They argue that "the struggle to improve educational attainment across the board and reduce the marked disparities in outcomes that are so troubling will take place mainly within the public universities" (p. 10). This is because two-thirds of all full-time students pursuing four-year degrees attend public institutions, be it a research university or a comprehensive four-year institution. The "struggle" in challenging the educational disparities is twofold. On one hand, students, particularly FG, LI students, must gle toward degree attainment in an environment unaccustomed to their growing presence on campus. On the other hand, four-year institutions, particularly public institutions, must struggle to find ways to serve their many roles and constituents while staying true to their main purpose: to serve the children of their state.

In theory, these flagship universities "were created to meet the social and economic needs of the states that chartered them, to serve as a great equalizer and preserver of an open, upwardly mobile society and to provide an 'uncommon education for the common man'" (Tobin, 2009, p. 240). Yet more and more, this goal has been shifted by specialized research foci, graduate instruction, grants, and consulting to the "gradual denigration of undergraduate education" (Tobin, 2009, p. 255). With shrinking funding from state budgets, flagships have

increased tuition and selectivity and have been appropriately criticized for giving up their role as the "people's university" (Tobin, 2009) to become the multiversity (Kerr, 1995). The changing demographics of this nation and the increasing presence of FG, LI students, students of color, and immigrant students demand that public universities, their faculty, staff, and administrators reconsider their collective role and mission in "advantaging the disadvantaged" (Levin, 2007, p. 185).

The purpose of this book was to deepen and add student perspectives to educators' understanding of the experience of FG students, particularly at four-year public institutions. Critical multicultural pedagogy enacted in learning communities was proposed as one vehicle to challenge the isolation and marginalization that many FG student experience on large, predominantly White college campuses. Students' narratives point out shortfalls in their university experience and also provide direction for what might be done to bridge these gaps. While there are no quick fixes, this chapter highlights the voices of the same FG students featured in the previous two chapters, with a focus on their critiques of higher education, and concludes with suggestions for how we in the academy might examine, shift, refine, and reconsider how we work together with this new majority of FG students.

In this chapter, I honed in on a few, critical quotes that capture sentiments shared by many of the LI, FG students interviewed in this study. Their observations speak to what they value in education and also what they see as problematic both locally and systemically.

YOU COME AS AN INDIVIDUAL, YOU LEAVE AS AN INDIVIDUAL

A central challenge of the LI, FG student experience has been the isolation and marginalization that these students experience on campus and in the classroom. This problem is shaped by both the norms and *modus operandi* of the campus, along with the multiple life roles that FG students juggle. Boyer (1990a) details the necessity for a renewal of community on campus and illustrates how the demise of community is rooted not only in the changing demographics of the student body but also in a "feeling that colleges are administratively and socially so divided that common purposes are blurred, or lost altogether" (p. 1). Zahara spoke to both the dilemma of diversity and the ways in which the structure and pedagogy of most classrooms continue to be individualistic and teacher focused. She said,

> I'm sorry, but it's just lecture; most professors don't have group projects or group anything. You come in as an individual; you leave out as an individual. You know what I mean? You turn in individual papers; you turn in individual assignments, quizzes, exams. There is really no interaction, like, What's your name? Who are you? Where are you from? And how were you raised, what do you think about life? You know, we don't know that about anybody. We just assume by maybe the clothes they wear, the way they wear their hair. Most of the time, I will never know if I walk past most of the people that I see in class out there because I wouldn't even know to say hi.

Zahara's words, "You come in as an individual; you leave out as an individual," fly in the face of what many colleges and universities espouse about their undergraduate experience. Boyer (1990a) suggests that "what is needed is a larger, more integrative vision of community in higher education, one that focuses not on the length of the time students spend on campus, but on the quality of the encounter, and relates not only to social activities, but to the classroom" (p. 7).

Paul compared his Multicultural Learning Community (MLC) experience to other experiences at the university and acknowledged the difficulty he had adjusting to the "traditional" classroom once he completed his first year:

> So, like, we in the [MLC] had a really good relationship. I appreciate it; there aren't many classes that I have been in that people know me in that way. That was different because the university structures curriculum in a way where there isn't much flexibility for teachers and professors—there was this sense of comfort zone in the MLC, and I really miss that. But at the same time, too, you have to realize that is not what life is about all the time. I mean, it is an experience, you got to go on. . . . But my second semester, I did miss it. I had to find a whole way of getting adjusted to the quote unquote "formal traditional education,

role of reln → unconditional (+) regard

standardized curriculum style," where everybody just goes to class and sits down and take notes and stuff like that.

Paul's reflection raises several issues about what he values in his learning and how he rationalizes his adjustment to the university. Like Zahara, he craves interaction and engagement with others and attributes this "comfort zone" to motivating and deepening his learning. He also acknowledges the ways in which instructors are constrained by the limitations of class size or scheduling that influence how they teach. Most troubling, though, is that he had come to expect that the "standard" college curriculum would be delivered to him by the faculty, while he and his peers passively received knowledge as external to them. He chalks up his learning community experience to an anomaly in his academic career, though he is constantly searching for places and people that could resonate with his need for connection and motivation. In the previous chapter, Paul talked about needing to be part of a larger collective, and he went on the address how feeling connected to his class, instructors, and peers sustained him. He reflected on his first two years of college in the MLC and as part of the University of Minnesota General College community:

> It's funny, ever since I left GC, my grades started dropping, and it's not because I can't do better. It's because there was no motivation. I am serious. There was no motivation. I have joined ROTC, and I am glad that the Air Force is already motivating me now. I have a very good relationship with my flight members, and they are expecting big things of me over there. And they really want me to do this, and the grades have to pull up. That is why I am working hard; that's my motivation. Ever since I left GC, there has not been any motivation.

Levin's (2007) extensive study of nontraditional students, including FG students, echoes Paul and Zahara's sentiments about the individualistic approach to being and learning on many large campuses. While there may be a myriad of programs, offices, and initiatives dedicated to serving these students, they are rarely embedded in the academic experience, which is where most of these students spend their time interacting with one another. Levin (2007) argues that students who have been on the margins and beyond need "guides and mentors . . . financial aid . . . a peer community . . . and academic integration that includes a plan to move them toward specific goals. These needs are met on an individual basis for a relatively small population of students; this is not a systemic approach taken by institutions or by state governments" (p. 191).

In many ways, the MLC was just that—an experience that met all of these four needs for a small group of students who completed their first year with a vision of learning in community that was incongruent with much of the rest of their academic journey. This is a critique not only of the university's approach to undergraduate education but also of the MLC itself. How effective is a curricular innovation that exists on the fringe of "traditional" modes of delivery in the classroom? This question will be discussed in more detail further on in this chapter.

Over There

One subject that emerged in the students' critiques of the academy was the extent to which they often felt discounted or invisible to faculty and staff. Some students were able to find and develop a few meaningful connections with instructors and advisors outside of the MLC. But many characterized the majority of their interactions with staff and teachers as dissatisfying. Zahara described her efforts to connect with an advisor in her major. She noted that this advisor was a graduate student who was not affiliated with her major and seemed more interested in "processing" her questions rather than getting to know her as a student. Zahara uses the phrase *over there* to describe her experiences in the larger university community. Her choice of words captured succinctly how she continued to feel like an outsider on campus:

> In the MLC, we were all willing to express what was going on in our lives, but over there—I call it *over there*—I didn't feel like anybody cared. I felt that people cared here, and so I think that when you feel people care; you feel this obligation to do your best and give your all, because you know they care about you. But when I got over to my new college, even now to this day, I don't feel that anybody cares. I even went to my advisor over here, and she's like, I don't care. I told her that I needed to meet with her and she said, "You need to do this and this," and that was over the email, and I'm like, OK . . . ? Let me know a time that would be convenient to meet, but no, she said, "This is what you need to do" over email. I am like, no community here. They don't do that *over there*.

Zahara's interpretation of this interaction reaffirms the ways in which students, particularly those who are new to the academy, need and value relationships with staff and draw upon these connections as sources of support and encouragement. Her expectations of the advisor clearly clashed with the perceptions that this staff person had of her role. In some cases, large advising loads force advisors to follow a model predicated on efficiency rather than on relationship. In other instances, advising is viewed as just that—providing specific advice on a given topic with little attention to the context or student posing the question.

Like Zahara, Anna addressed how many of her exchanges with faculty suggest that they were unable or unwilling to engage with her as an individual but wanted to keep her at a distance, as one face in a sea of many: "I just think about the relationship between the professor and student. . . . Professors don't really care who you are, what your name is; they just want that paper to give you your grade, you know? Yeah, I know that that is the case sometimes, but sometimes you want to know somebody on a more personal level, you know? Like, there is more to it than that, there is more to me than just, a number—my student ID number."

In her need to be recognized as a multidimensional learner rather than as a face in the classroom, Anna expressed integration between her interpersonal and cognitive self and her need for a classroom experience that understood this intersection. Hooks (1994) writes about recognizing this "desperate need in students—their fear that no one really cares whether they learn or develop intellectually" (p. 203). Like Anna, many FG students seek to engage in knowing that is relational and connected; as such, they need teachers who are willing to share the "process and the product of their thinking" (Clinchy, 2000, p. 33).

Jarod's comments echoed Anna's dissatisfaction and dug deeper into the hierarchical nature of higher education, where rank, title, and position dictate relationships between faculty and students:

> There are these professors, and they have, like, a kind of higher ranking, so therefore they're smarter than I am. But it's the professors who didn't take that train of thought who really felt like "this person is, you know, is not my just my student, but they're actually a human." [Laughs] Teachers like that are kind of the ones that I felt the most connected to. But I kind of feel like a lot of teachers, have this . . . [Pauses] I don't think it's blatant; I think a lot of it is really subtle. There's like this idea that because you have a PhD that you're better than somebody.

Jarod points to subtle, covert, and unsaid ways in which the element of power is woven into the teacher-student relationship. Some teachers are able to shift this power and engage in "*collateral learning*" (Rendón, 2009, p. 96), where students and teachers are co-creators of knowledge. Yet in many cases, the culture of higher education perpetuates these hierarchies and values disciplinary expertise over lived experie than considering the interdependency of these dual ways of knowing. Each student above expressed their frustration and perception that most faculty do not care about them as multidimensional individuals. Yet we know that many faculty care deeply about students and are troubled by what they perceive as students' lack of engagement or by their own inability to connect with students (Rice, 1996).

A good part of this problem can be attributed to the fact that the academy and graduate school do little to encourage faculty to see their teaching as an integrative practice. More often than not, faculty and the tenure-track process itself perpetuate the very dualistic stances that we try to encourage our students to eschew. Teaching and learning are seen as distinct and separate from the students' experience and personhood. Hooks (1994) writes that "teaching is often the least valued of our many professional tasks . . . and given that our educational institutions are so deeply invested in a banking system, teachers are more rewarded when we do not teach against the grain" (p. 203).

Many faculty have little experience or training in working with diverse student groups and considering the context of the new majority in their classrooms (Anderson, Gaff, & Pruitt-Logan, 1997; Gaff & Lambert, 1996). Rendón (2009) suggests that we pay close attention to "faculty positionality" and reexamine our role in the classroom from the points of view of our students. To begin this process of self-reflection as teachers, we must begin to deconstruct how the academy and our own disciplinary training views knowledge and start reevaluating how to contextualize our teaching in order to bridge the gap between students as learners and students as co-learners.

We're Going to Teach You How to Be a Scholar

As more and more FG students find their way to into public, four-year institutions, we will have to learn to be more honest about the ways in which the academy favors certain ways of knowing to the detriment of others. We will have to confront the explicit and implicit ways in which we are limited by the parameters of our own disciplinary training and acknowledge how well we have been trained to discount and discredit approaches and ideas that do not fit into our own authoritative text. In doing so, we can begin to "confront higher education at the core, engaging the politics of knowledge and the historically defined structures that privilege the elite at the expense of the masses" (Osie-Kofi, Richards, & Smith, 2004, p. 56).

Students—*all* students—need to understand the key concepts in disciplines and courses—this is not in question. Nor is the reality that no amount of rich, lived experience alone can prepare FG students or any students to become proficient in their fields of study or their careers. What is at issue is the manner in which the academy *communicates* this expertise to students and "privileges intellectualism at the expense of inner knowing" (Rendón, 2007, p. 112). In addition, more and more, we see that "a deep understanding of contemporary life requires knowledge and thinking skills that transcend traditional disciplines" (Boix Mansilla, 2005, p. 15) and engage students in multiple ways of knowing (Anderson et al., 1997).

Jarod described how faculty who viewed knowledge as "bounded collections of facts" (Boix Mansilla, 2005, p. 16) confined, constrained, and distanced him from the academy:

> It was just a constant "You have to adhere to this way of thinking." And that's kind of what academia is. That's how I perceived it: "You're telling me my way of looking at this is wrong, and I need to adjust it to this way of thinking. And this way of writing it is wrong. I need to be able to communicate this to scholars." And scholars won't understand if you write it this way. In the university—there wasn't that dialogue that we had in the MLC. It was like, "This is how it is, and this is how we interpret this theory. And interpret this data, and this is how it is." Period, point, blank. And you know, "I'm the PhD, so I know, and you're just the college student, so your ideas aren't important, *yet*." So that, and that's how I felt a lot of times.

He went on to say that while he did not expect that his contributions would always be insightful or correct, he appreciated and valued the opportunity to have input and to process his thinking toward forming an answer: "And so the difference between the MLC and being in my liberal arts environment was like, even if I was wrong about something or I might have misinterpreted something, the fact that I was taking it all in was a good thing."

Jarod's insights represent how many LI, FG students view the academy and walk the line between trying to acculturate to the expectations and norms without losing themselves in the process. The result of this struggle can give rise to a sense of bitterness, of having to give up too much in order to be absorbed into this conventional community of scholars. Wilson (1995), a working class academic wrote, "The whole process of becoming educated was for me a process of losing faith" (p. 214). This sentiment, expressed by both Wilson and

Jarod, reflects the many contradictions that LI, FG stud
entering and surviving the academy. On the one hand,
the value of the degree toward social and economic mo
and their family. On the other hand, they must reconcile this reward with experiences that often challenge their multiple identities, histories, and home worlds. Reconciling one's place in higher education often prompts LI, FG students to consider their opportunity to attend college in the context of their community and family.

The following section explores the push-and-pull dynamic of these students' relationships with and ultimate accountability to both their academic lives and communities. In addition, students grapple with the extent to which public education in both secondary and higher education takes accountability for the success and failure of historically marginalized students.

Somewhere We Have Lost Something

As the last section addressed, many LI, FG students recognize that in coming to college, they must give up parts of their old selves, and with varying degrees of success, some begin to develop a new sense of self-authorship that honors their many identities. There is still a sense of loss and also a sense of deep responsibility toward those who will walk these same roads after they have graduated. This sense of responsibility to serve as a conduit to their communities is coupled with an understanding of the many stakeholders who shaped the destiny of the next generation.

Ruben addressed how his own experiences and challenges getting to and surviving the university encouraged him to engage in community service to bring young, disadvantaged male students to campus in an effort to plant a seed—to allow them to see that college was a possibility for them. In doing so, he noted that the university, despite its land-grant mission and its geographical proximity to many low-income neighborhoods and communities of color, was still a closed and unknown space to these young students:

> I wanted to go out to the community to cultivate the young. Yeah, we did a lot of tutoring and mentoring, and we brought a lot of kids from Minneapolis to the university. What was absolutely mind boggling to me was that they lived so close to the university—like, literally blocks away—and had never been here, which is crazy. Most of them were disadvantaged youth; they had never been to the university, and we would bring them to the classroom and discuss what it means to be a

> male, what it means to, you know, to date a girl . . . putting them in that classroom is the first step in planting that seed—that idea that college is a possibility.

His observations point to the growing selectivity of many public research institutions, where the percentage of enrolled LI students dropped from 14 percent in 1995 to 9 percent in 2003. In a Lumina-funded study that examined the growing inequities at public flagship universities, Gerald and Haycock (2006) document that while high school graduating classes are increasingly diverse, this diversity is not equitably represented in the freshman classes at public institutions. Their data analysis suggests that "accessibility of these institutions is declining, success rates of minority students are lagging, and most institutions are getting worse, not better" (p. 8).

Zahara commented on these growing gaps of opportunity between the haves and have-nots when she reflected on the current condition of the inner-city high school from which she graduated. Her frustration with the public school system, the university, and the condition of her community was evident in her remarks:

> Yeah, maybe we might have low test scores . . . but where did these kids come from before they came to North High School? There are all these other schools that these kids come from. The test scores are low, what are *you* doing to increase the test scores? What are you doing to put good teachers that are going to teach our students what they need to be taught? What are you doing? Because I think we have lost that somewhere, you know what I mean? Like somewhere in the system, we have lost a lot of money over there. Somewhere we have lost something. There is no real love or no real pride in where we are going or what we are going to do as far as an inner-city community because there is nobody there . . . you know, it's getting worse, because the students, the younger kids aren't coming in like we did. You know what I mean?

Zahara's comment hones in on the necessity for partnerships between flagship institutions and school systems, particularly in communities that have historically been underserved in education. Public land-grant institutions are beset with their own funding woes and dwindling support from the state and cannot fix all problems in public education. But the move away from access and toward selectivity is not the answer: "Universities themselves are important actors in this drama of shrinking opportunity. Not victims, not side-line spectators, but independent actors" (Gerald & Haycock, 2006, p. 4).

Shifting the Center

For close to two decades, I have been immersed in teaching, advising, and working with LI, FG college students. Teaching in the MLC allowed me to know and see these students in new ways, and when the semester was over, I became deeply immersed in reading, rereading, and analyzing their writing narratives in hopes of telling the story of their experiences in higher education. When I received funding to interview students three to four years after the MLC experience, I was able to add another chapter to this journey and consider how they made sense of their college experience beyond the first year. My primary goal was always to showcase their voices, their interpretations, and their perceptions of college so that we could reimagine the academy in ways that honored, welcomed, and valued them. As I weigh their positive and negative experiences, I am heartened by their resilience, insights, and successes, but I am also acutely aware of the ongoing dilemmas that continue to cloud the way of others like them. Years ago, I attended a talk by a University of Minnesota Professor Emerita Toni McNaron, who spoke about the immense challenges in trying to create paradigm shifts in higher education—particularly around issues of diversity, inclusion, and the status quo. Using the metaphor of a mobile, she suggested that "while the academy supported many diversity initiatives both programmatic and curricular, they are represented on the fringe of a mobile, like satellites that orbit the core of the structural framework. As a result, the smaller satellites may spin furiously, but the nexus of the academy is unchanged by their movement. However, any change in the core of the mobile creates huge shifts in the satellites" (McNaron, as cited in Jehangir, 2009b, p. 285).

I find this metaphor to be particularly relevant to the critiques and the narratives of the FG students featured throughout this book. Indeed, the core dilemma of the MLC was the way in which it was one such small satellite, and while it sustained and propelled many students to success in college, it also failed others. Even for those who succeeded in graduating, the university often remained a place *over there*. The MLC was an experiment in a *learning community*. It drew upon interdisciplinary ways of knowing, lived experience, and critical multicultural pedagogy as a political stance about reimagining how we might come together as co-learners. In some cases, students were able to translate these experiences into how they viewed themselves and how they navigated the academy.

But many students also felt that leaving the MLC was like starting all over again, only with the knowledge that what they had come to value as learners was unlikely to be experienced again in the "traditional curriculum." This dichotomy suggests that if the academy is committed to inclusivity, then we need to find ways to shift our focus and move the center of the proverbial mobile to reshape the experience of LI, FG, and all students on campuses. Recent changes in the academy, specifically at many public research universities, have begun to shift the nexus of the mobile to impact LI, FG, and in fact, all students in both positive and negative ways. These changes include not only increased selectivity at land-grant universities but also an increasing awareness of the need to reexamine the role and delivery of undergraduate education, particularly in lieu of the changing demographics in the student body (Association of American Colleges and Universities [AACU], 2007; Bauman, Bustillos, Bensimon, Brown, & Bartee, 2005, 2007; Boyer, 1990b). As a result, there appears to be a window of opportunity for us in the academy to ask ourselves some big questions about how much we are willing to shift our old ways to challenge the "jarring disconnect between aspiration and actual achievement" (AACU, 2007, p. 7) of many historically marginalized students. What can we do on our campuses, in our classrooms, and in our policies to give these students "full citizenship" on campus?

Faculty as Creators of Change

While many students in this study were frustrated and dissatisfied with their experience with faculty, they also found a handful of teachers who became their guides and mentors throughout college. In some cases, this interaction may have been short lived but powerful enough to inspire a student to study further in a given subject area. Those who were able to cultivate these relationships, both in and outside of the classroom, were supported academically and developmentally by their teacher mentors. These particular faculty members were able to bridge the gap between student development and student learning to "a more holistic and integrative approach to college learning" (AACU, 2007, p. 18). So why aren't more of us doing the same? To begin with, graduate schools do a poor job of helping future faculty members envision themselves as teachers. With the focus on research and disciplinary expertise, few faculty arrive in the classroom having considered the many issues that will shape their curricular, pedagogical, and philosophical approach to teaching.

In his work, *Scholarship Reconsidered*, Boyer (1990b) argues that "what's needed is a requirement that teaching training be incorporated into all graduate preparation" (p. 71). Such a requirement would demonstrate a true commitment to the scholarship of teaching. Even the faculty at research universities who are committed to teachir caught between the competing obligations of teaching and research, where more than half agree that "the pressure to publish reduces the quality of teaching" (p. 55). The tenure process itself is a self-focused journey; it does little to encourage the new professoriate to consider innovative approaches to teaching, to link service engagement to research, or to engage in community partnerships. All such endeavors are often viewed as unpredictable projects by many in the academy and are considered to take the focus off research and publication, as if there could be no linkage between the aforementioned activities and research. Consequently, the tenure process and much of the traditional role of the professorate is deeply individualistic, wherein the rewards "emphasize individual rather than collaborative excellence in both scholarship and teaching, which results in system disincentives for faculty members to spend their time in collaborative redesign of undergraduate education" (AACU, 2007, p. 47).

Yet we are at a pivotal time when the necessity to reexamine undergraduate education at public four-year institutions is not only the right thing to do but also an economic necessity. Simply changing admissions requirements to avoid addressing the gaps between high school preparation and college expectations is a short-sighted response to the demands of an increasingly diverse student body and society. How, then, do we as educators make the shift to support, engage, and learn from the LI, FG students who arrive in college classrooms? Regardless of programmatic and co-curricular organizations created to support LI, FG students, "faculty and teachers who work directly with students are the only ones who can make it happen" (AACU, 2007, p. 48).

One approach might be for us, as teachers, to grapple with the same issues that we ask our students to take on. The American Association of Colleges and Universities (AACU) has suggested that college learning needs to adapt to a new global century in order to achieve equitable educational outcomes from all students (2005, 2007). One recommendation from these reports is for administrators and teachers to consider their role and responsibility in challenging the achievement gap on campus and beyond. A critical step in doing this—particularly at large, fragmented, multidisciplinary research universities—is a campus dialogue across and among stakeholders that invites participation from the various branches or factions of the organization. If there is to be renewed commitment to undergraduate education and to the new majority of students in our classrooms, it cannot be piecemeal.

Currently, depending on the major, department, or campus, a select group of faculty may been seen as "responsible" for undergraduates, and that group gets further divided into those who are working with first- and second-year students versus those who work with juniors and seniors. Graduate faculty in the same department or college may have little or no overlap with these roles and may see themselves as removed from undergraduate education. It is appropriate for each of these groups to have different and specific areas of focus. However, these groups' failure to see the connection between their collective work trickles down to undergraduate students and especially to LI, FG students on campus. Graduate faculty may not teach undergraduates, but they prepare other faculty to work with undergraduates. Their responsibility lies in revamping the graduate curriculum to prepare future faculty for a life in the classroom and to better understand the diverse students with whom they will engage there. As for those who work with undergraduate students directly, our work is more complicated and asks us to consider which students we can truly see as "ours," belonging to and composing the academy.

In the last two decades, teaching in college has undergone drastic changes in delivery, pedagogy, and philosophy. Despite initial resistance, multicultural curriculum, interdisciplinary approaches to learning, active teaching and learning pedagogies, and learning communities have become more and more prevalent in college classrooms across the country (Kuh, Kinzie, Schuh, & Whitt 2005). At many large public institutions, however, these innovations are still outside the mainstream and as such, remain as "isolated initiatives" (Williams, Berger, & McClendon, 2005, p. 2) with limited capacity to have long-reaching effects on the students they serve. Those who chose to participate in these collaborative efforts are driven by their personal values and a commitment to enhance undergraduate education but are often "on the edges of political and social power structures" in the academy (Smith, McGregor, Matthews, & Gabelnick, 2004, p. 242).

Higher education is slow to change, and it would be difficult to "envision a campus where inclusion is a necessary factor in achieving excellence but where faculty were not involved in the change process" (Williams et al., 2005, p. 15). Senior faculty can play a major role by advocating change in the classroom and by reexamining their methods of teaching in undergraduate education. Engaging in this process involves considerable personal and professional risk; it involves challenging historic and symbolic ideas about the role and function of faculty, as four-year and research universities work to find a better balance between teaching and research. Perhaps these goals can ultimately be obtained by rewarding teaching innovations in the same manner as publication success. Faculty development opportunities can deepen our understanding of LI, FG students, enabling us to see intersections among their multiple identities and to consider affective and cognitive ways of knowing, as well as the strengths that diverse cultural learners bring to the academy (Williams et al., 2005). We can also draw on the expertise of colleagues on our campus to create teaching communities that provide us with the *space* and *place* to discuss, strategize, or process challenges and successes in teaching LI, FG, and all students.

Creating Coalitions

Another step toward shifting how we view LI, FG students, and indeed, all undergraduate students on campus is to connect with others on campus who see them in different contexts. Building meaningful coalitions between faculty and student-affairs professionals will be a critical part of creating webs of support as students learn to navigate the academy (Nash, 2009). As is the case with disciplinary borders, there have also been historical boundaries between faculty and student-affairs professionals, particularly at large public schools where there is an unspoken agreement that the faculty is concerned with matters of the mind while student affairs is relegated to all other issues pertaining to student development.

Despite these disparate roles, it has been well documented that students, especially historically marginalized students, are most engaged on campuses where a supportive environment, coupled with active and collaborative learning and faculty interaction, allows them to navigate academic challenges (Kuh, 2003, 2008; Kuh et al., 2005). These connections become more fluid and deep on campuses where faculty and student-affairs professionals work in direct partnership to meet students' needs, both inside and outside of the classroom (Kinzie & Kuh, 2004). Nash (2009) argues that faculty can learn a great deal from student-affairs practitioners because of their proximity to different aspects of campus life—residential living, student organizations, advising, counseling, and more. Similarly, advisors can get a deeper understanding of the classroom context, challenges, and students' behaviors by connecting with faculty. On some campuses, faculty and student-affairs professionals share classroom space; in other instances, faculty might serve as advisors for a student organization or be involved in planning orientations.

These opportunities to develop crossover roles must be encouraged and rewarded because they are an effective means of crossing boundaries and dissipating hierarchical structures that are detrimental to both staff and students on campuses. By knowing and respecting one another, university staffs are able to better serve and understand the needs of diverse students. Consider the following scenario: An LI, FG student misses a week of class and does not respond to emails sent out by a concerned faculty member. The following week, she returns to class, and through conversations with both a faculty member and an advisor, she reveals that she was absent because her family's home was foreclosed and that she spent the week helping her family pack up their belongings for the move. The partnership between faculty, advisor, and student created the opportunity to provide the student with support at a critical juncture in her life and also helped to manage the planning it would take to get the student back on track and complete her semester in college. Coalitions between the classroom and the campus can occur in multiple ways. Effective coalitions are not about which program or staff person has more claim to a student but rather about how coalitions help us all better understand student needs and strengths to create "mentoring communities" (Parks, 2000, p. 134) for LI, FG students.

On large campuses, there are many programs that serve many different student needs, and in the case of FG students, this gets more complicated because this population of students is not homogenous. In fact, they have multiple identities that might be well served by a student-parent support group, disability services, or any number of student organizations. This is one of the strengths of the large campus. A way to increase this strength is to find ways to link organizations to the academic lives of students and to resist the temptation to create more and more new programs. What we need is better communication among the many representatives who serve LI, FG students on campus. One place to begin would be to look to TRIO programs on our own campuses that not only provide much needed external funding to work with LI, FG students but are also experienced in advocating for the needs and strengths of this particular student population. Often LI, FG students are required to work with numerous offices and programs, and even though these programs each start out with good intentions, they often duplicate each other's roles or work at cross-purposes. Assessing LI, FG students' needs and creating networks between existing campus programs, from financial aid to advising to scholarship programs, are another means of creating coalitions than serve rather than stress our students.

Minding the Gaps

There are two critical issues that continue to challenge even the LI, FG students who have a strong start to college in the MLC: the first

involves surviving the transition from the first to second ye
lege, and the second involves integrating into the larger sc
the university. The transition that students must make between their first and second year of college is a period that has been marked as problematic for many students who get lost or become overwhelmed by decisions about majors, loneliness, and dissatisfaction with college (Hunter et al., 2009). As has already been discussed, LI, FG, students leave college after the first year more often than their more traditional peers, thus reinforcing the need to maintain support systems that sustain them. In this study, students who developed a sense of belonging in the learning community felt even more alone once they left the community. While many did not drop out, they still described a lack of motivation and difficulty adjusting to the "traditional curriculum." Many missed their MLC cohort but did not have an organized means of staying in touch, as the demands of new roles, classes, and schedules took precedence. These concerns reflect the gaps between a small-scale innovation and relatively stagnant landscape of the larger university. While incremental steps by faculty, staff, and administrators may create larger, more holistic changes in undergraduate education, the need to immediately capitalize on current innovations is critical.

For campuses that have learning communities and programmatic support for LI, FG students, it is vital to build bridges between these first-year programs and the second year of college. These bridges can range from facilitated reunion groups that allow student cohorts to stay connected, to sophomore academic offerings that allow students to build on the identity development and belonging they established in their first year. One- to two-credit seminars that allow students to consider the nature of the transition from first to second year, allow for career and major exploration, and continue to maintain a regular check-in with a cohort of peers may be one such bridge. In addition, LI, FG students who have completed their first year can serve as mentors to new first year students, a role that reinforces their own role as teachers and allows them to share how they learned to navigate the particular challenges of their institutions. The specific design of academic and programmatic links in the second year of college can best be determined by the specific needs of LI, FG students at a particular institution—the key is to prevent undoing the work of successful first-year innovations by failing to support students into their second year.

The second major dilemma that LI, FG experience involves the issue of integration versus separation—it is critical that FG students do not merely exist on the boundaries of campus but are instead able to see themselves as integrated full citizens of the university. For this to happen, they must be able to feel a sense of comfort, belonging, and engagement with all students in many different settings. This is, indeed, the ideal, and multicultural scholars have underscored that when diverse groups engage with one another, opportunities for critical reflection, prejudice reduction, and a deeper understanding of multiple perspectives are far more effective than simply reading about these issues (Chang, 2002; Chizhik & Chizhik, 2002; Hornak & Ortiz, 2004).

While I agree wholeheartedly with this assertion, I would also argue that FG students need opportunities to engage with one another, to cultivate their voice and personhood in the academy. They need spaces and places where they have a *critical mass*—where they do not have to explain themselves or mask their experiences to blend into college. FG students are not a homogenous group, as the student narratives in this book reflect; they are extremely diverse, yet their common experience as the first in their family to come to college *does* bind them together.

In the previous chapter, themes of belonging and identity emphasized how the peer group in the MLC had a deep impact on the students' comfort level in the classroom, their willingness to share their own lived experiences, and their ability to utilize disagreement as a springboard for knowledge construction Having safe spaces to make sense of their social and academic identity development is a powerful precursor to their ability to engage meaningfully with *all* students in their college experience. With attention to all we know about the challenges that LI, FG students face in college, it would be of great importance to consider how to create spaces specific to them, as well as spaces that allow them to engage and connect with students across campus.

Yours, Mine, and Ours: Education for Democracy and Inclusivity

Despite the ideals of equality that initially forged American education, it has in many ways become a class system where the nature of one's opportunity depends on ascription. Inequalities surrounding class, race, gender, and other aspects have funneled some students into better opportunities than others. As the demographics of this country have changed, FG students have been, in some ways, the "gate crashers" at four-year colleges. Institutions across the nation are being forced to take note of these changes in their student body, and some are choosing to make intentional efforts to retain and support these students (Kuh et al., 2005).

As Peter Sacks (2007) notes in his eye-opening book, *Tearing Down the Gates*, "It is a revolution of sorts when major universities decide to push back against some of the powerful and entrenched forces that have helped to create these inequities. A few of America's best public universities are trying to find fairer more complete, and even more accurate ways of measuring merit" (p. 225). As access to college opens up the gates to students who change the feel of our classrooms and campuses, we all bear equal responsibility in shifting their status from one of gate crasher to one of full citizen.

To make this shift, we will need to break from our previous views about whom we are responsible to and what we are responsible for. Within large, flagship universities, we serve students with a range of interests, challenges, and talents. As long as students who are LI, FG remain on the periphery of our collective responsibility, they will not succeed on our campuses. As long as we remain tethered to the idea that our work is with only with science students, or honors students, or students within a specific major, we fail to see the potential of LI, FG newcomers joining us in the academy. Students who come to our institutions are *our* students—all of them.

Part of shifting how we view LI, FG students involves reframing the deficiency lens to consider their strengths. While many of these students may need pedagogy that supports academic skill building, they also have experience adapting to new environments, languages, and cultures that would astound their more traditional peers. The stories I have shared in this book are a testament to these students' many roles as cultural brokers, caretakers, students, employees, and parents. We as educators play a role in helping them claim these strengths and translate their lived experience into skills that they can use to navigate college.

We as educators cannot be expected to have all the answers, and we may not even be asking all the right questions. This is why it will be important for us to engage LI, FG students in the process of shaping their own destiny in higher education. Too often in the academy, "there is a tendency to assume a problem is understood and to come up with solutions that may do nothing to address it" (Williams et al., 2005, p. 13). Students are experts on their experience on campuses, their insights about campus climate, programmatic support, and academic experience, and we would be wise to include them in dialogues about reforming undergraduate education.

This book has been an effort to begin such a dialogue, to situate our discussions in the stories, perspectives, and lived experience of FG students, and to raise questions that we might want to try answering together with the help of students. It is a call to consider how teaching and learning in higher education can be a means of practicing educational reformation that challenges institutional impediments that limit who succeeds in college.

CLOSING REFLECTION

If you are lucky enough to work on a college campus in America, you have undoubtedly walked across campus sometime in the early fall. The classes have just begun, and students laden with backpacks are making their way through campus into their educational journey. I never tire of being part of that liftoff, that feeling of possibility in starting afresh or starting again. I sense that same energy and excitement in the students I work with, even if it is intertwined with a healthy dose of anxiety. That energy is about the possibility and promise of education in all its forms. Those of us who are privileged enough to work in higher education are bound to consider how we might capitalize on the hopes of our students and the hopes of their families and communities.

For all students, especially LI, FG students, making good on the promise of college education is to reexamine how we teach, what we teach, and who we teach. It is to recognize that these students' failures are not theirs alone but also tied to us as educators. Making substantive changes in how higher education invests in historically underrepresented students is critical, not only to the individual lives of our students but also to the social and economic health of our larger society. Our commitment to serving a new majority of students is about them and us and the bigger picture. As Paul said,

> I was able to realize a bigger picture, behind not just the frustration, but to just identify the group goal. 'Cause, like, I can get frustrated and say, "You know what, I don't want to be a part of this" and walk away. But at the end of the day when you think about it, it's not just about you, it's about the group. So when you think about the collectiveness of the group, you become invisible in some ways because you think about the aim of the group—it's not about you—it's about what we *have* to do. So like, at some point in time, like, I have learned in life, not everybody is going to buy your product, but at the same time, too, if you think about it, what is the ultimate goal of the group? To see the ultimate goal of the group, you have to ignore some of your feelings and think about what everybody else is doing, and if this is addressing the aim of the group—that is what you need to be supporting, not just what you think.

Should we take on the challenge of answering these questions, we, like our students, will be forced to consider new ideas, new ways of working together, and new ways of seeing ourselves. We will be uncomfortable, and we will be frustrated. This is progress.

References

Abes, E. S., Jones, S. R., & McEwen, M. K. (2007). Re-conceptualizing the model of multiple dimensions of identity: The role of meaning-making capacity in the construction of multiple identities. *College Student Development, 48*(1), 1–22.

Adelman, C. (1999). *Answers in the tool box: Academic intensity, attendance patterns, and bachelor's degree attainment.* Washington, DC: U.S. Department of Education, Office of Educational Research and Improvement.

Adelman, C. (2007). Do we really have a college access problem? *Change, 39*(4), 48–51.

Alexander, C. A, Ishikawa, S., Silverstein, M., Jacobson, M., Fiskdahl-King, I., & Angel, S. (1977). *A pattern language: Towns, buildings and construction.* New York: Oxford University Press.

Altschuler, G. C., & Kramnick, I. (1999). A better idea has replaced "in loco parentis". *Chronicle of Higher Education, 46*(11).

Anderson, H., Gaff, J., & Pruitt-Logan, A. S. (1997). *Frequently asked questions about preparing future faculty* (Vol. 3). Westport, CT: Greenwood Press.

Anzaldúa, G. (1999). *Borderlands: La frontera* (2nd ed.). San Francisco: Aunt Lute Books.

Apple, M. W. (1990). *Ideology and culture* (2nd ed.). New York: Routledge.

Archer, J. J., & Lamnin, A. (1985). An investigation of personal and academic stressors on college campuses. *Journal of College Personnel, 26*, 210–215.

Association of American Colleges and Universities (AACU). (2007). *College learning for the new global century.* Washington, DC: The National Leadership Council for Liberal Education and America's Promise.

Association of American Colleges and Universities (AACU) National Panel. (2002). *Greater expectations: A new vision of learning as a nation goes to college.* Washington, DC: Author.

Association of American Colleges and Universities and the Carnegie Foundation for the Advancement of Teaching. (2004). *A statement on integrative learning.* Washington DC: Author. Retrieved from http://www.aacu.org/integrative_learning/pdfs/ILP_Statement.pdf

Astin, A. W. (1994). Student involvement: A developmental theory for higher education. *Journal of College Student Personnel, 25*, 297–308.

Astin, A. W. (1995). *Achieving educational excellence.* San Francisco: Jossey-Bass.

Astin, A. W. (1997). *What matters in college: Four critical years revisited.* San Francisco: Jossey-Bass.

Banks, J. A. (1981). *Multicultural education: Theory and practice.* Boston: Allyn and Bacon.

Banks, J. A. (1992). Multicultural education: For freedom's sake. *Educational Leadership, 49*(4), 32–36.

Banks, J. A. (2001a). Multicultural education: Characteristics and goals. In J. A. Banks & C. A. Banks (Eds.), *Multicultural education issues and perspectives* (4th ed. pp. 3–26). New York: Wiley and Sons.

Banks, J. A. (2001b). Multicultural education: Historical development, dimensions, and practice. In J. A. Banks & C. A. Banks (Eds.), *Handbook of research on multicultural education* (pp. 3–24). San Francisco: Jossey-Bass.

Barefoot, B. (2000). The first-year experience: Are we making it any better? *About Campus, 6*(4), 12–18.

Barefoot, B. (2009). Engaging Today's first-year students: Challenges and opportunities in the classroom. Key note speaker at Realizing Student Potential I-Teach Conference sponsored by Minnesota State and Community Colleges. February 28th 2009. http://www.ctl.mnscu.edu/events/rspconf/2008/speakers.html

Barr, R., & Tagg, J. (1995). From teaching to learning: A new paradigm for undergraduate education. *Change, 27*(6), 13–25.

Bauman G. L., Bustillos, L. T., Bensimon, E. M., Brown II, C., & Bartee, R. D. (2005). *Achieving equitable educational outcomes with all students: The institution's role and responsibilities.* Washington, DC: Association of American Colleges and Universities.

Baxter Magolda, M. B. (1992). *Knowing and reasoning in college.* San Francisco: Jossey-Bass.

Baxter Magolda, M. B. (1999). *Creating contexts for learning and self-authorship: Constructive-developmental pedagogy.* Nashville, TN: Vanderbilt University Press.

Baxter Magolda, M. B. (2001). *Making their own way: Narratives for transforming higher education to promote self-development.* Sterling, VA: Stylus.

Baxter Magolda, M. B. (2004). Self-authorship as the common goal of twenty first century education. In P. M. King & M. B. Baxter Magolda (Eds.), *Learning partnerships: Theory and models of practice to educate for self-authorship* (pp. 1–36). Sterling, VA: Stylus.

Baxter Magolda, M. B. (2007). Self-authorship: The foundation for twenty-first century education. *New Directions for Teaching and Learning,* (109), 69–81.

Baxter Magolda, M. B. (2008). Three elements of self-authorship. *Journal of College Student Development, 49*(4), 269–284.

Belenky, M. F., Clinchy, B. M., Goldberger, N. R., & Tarule, J. M. (1986). *Women's ways of knowing*. New York: Basic Books.

Bensimon, E. M. (1994). Bilingual cash machines, multicultural campuses, and communities of difference. *Innovative Higher Education, 19*(1), 23–32.

Bensimon, E. M., Neumann, A., & Birnbaum, R. (1989). Making sense of administrative leadership: The "L" in higher education. In *ASHE-ERIC Higher Education Report* (Vol. 1). Washington DC: George Washington University.

Bettinger, E., & Long, B. (2005). *Addressing the needs of under-prepared students in higher education: Does college remediation work?* Cambridge, MA: National Bureau of Economic Research.

Billson, J. M., & Terry, M. B. (1982). In search of the silken purse: Factors in attrition among first-generation students. *College and University, 58*(1), 57–75.

Blanton, L. (1999). Classroom instruction and language minority students: On teaching to "smarter" readers and writers. In L. Harklau, K. M. Losey, & M. Siegal (Eds.), *Generation 1.5 meets college composition: Issues in the teaching of writing to U.S. educated learners of ESL.* Mahwah, NJ: Lawrence Erlbaum Associates.

Boix Mansilla, V. (2005). Assessing student work at disciplinary crossroads. *Change, 37*(1), 14–21.

Bowen, W. G., Chingos, M. M., & McPherson, M. S. (2009). *Crossing the finish line: Completing college at America's public universities.* Princeton, NJ: Princeton University Press.

Boyer, E. L. (1990a). *Campus life: In search of community.* Princeton, NJ: The Carnegie Foundation for the Advancement of Teaching.

Boyer, E. L. (1990b). *Scholarship reconsidered: Priorities of the professoriate.* New York: The Carnegie Foundation for the Advancement of Teaching.

Boyer, E. L. (1996). The scholarship of engagement. *Journal of Public Service and Outreach, 1*(1), 11–20.

Braxton, J. M., Sullivan, A., & Johnson, R. M. (1997). Appraising Tinto's theory college student department. In J. Smart (Ed.), *Higher education: Handbook of theory and research* (Vol. 12, pp. 107–164). New York: Agathon Press.

Breneman, D. W., & Nelson, S. C. (1981). The future of community colleges. *Change, 13*(5), 17–25.

Britzman, D. P. (1991). *Practice makes practice: A critical study of learning to teach.* Albany: State University of New York Press.

Brookfield, S. (1995). *Becoming a critically reflective teacher.* San Francisco: Jossey-Bass.

Brubacher, J. S., & Rudy, W. (1997). *Higher education in transition* (4th ed.). New Brunswick, NJ: Transaction.

Bruch, P. (2002). *1422 Writing Lab.* Spring class syllabus, University of Minnesota, General College, Minneapolis.

Bruch, P., Jehangir, R., Lundell, D. B., Higbee, J. L., & Miksch, K. L. (2005). Communicating across differences: Toward a multicultural approach to institutional transformation. *Innovative Higher Education, 29*(3), 195–208.

Bruffee, K. A. (1999). *Collaborative learning: Higher education, interdependence, and the authority of knowledge.* Baltimore, MD: Johns Hopkins University Press.

Bui, K. V. T. (2002). First-generation college students at a four-year university: Background characteristics, reasons for pursuing higher education, and first-year experiences. *College Student Journal, 36*(1) 3–11.

Buriel, R., Pérez, W., De Ment, T. L., Chavez, D. V., & Moran, V. R. (1998). The relationship of language brokering to academic performance, biculturalism, and self-efficacy among Latino adolescents. *Hispanic Journal of Behavioral Sciences, 20*(3), 283–297.

Cabrera, A. F., Burkum, K. R., & La Nasa, S. M. (2005). Pathways of a four-year degree: Determinants of transfer and degree completion. In A. Seidman (Ed.), *College students retention: A formula for student success.* Westport, CT: Praeger.

Cabrera, A., & LaNasa, S. M., & Burkum, K. (2001). *Pathways to a Four-Year Degree. The Higher Education Story of One Generation.* University Park, PA: Center for the Study of Higher Education at The Pennsylvania State University.

Cabrera, A. F., Nora, A., & Castaneda, M. B. (1992). The role of finances in the persistence process: A structural model. *Research in Higher Education, 33*(5), 571–573.

Cabrera, A. F., Terenzini, P. T., & Bernal, E. M. (1999). *America's low-income students: Their characteristics, experiences and outcomes.* Report to the College Board. University Park, PA: Center for the Study of Higher Education.

Calcagno, J., & Long., B. (2008). *The impact of post-secondary remediation using a regression discontinuity approach: Addressing endogenous sorting and noncompliance.* Working paper, Community College Research Center, Columbia University, Teachers College, New York, NY.

Carey, K. (2005). *Choosing to improve: Voices from colleges and universities with better graduation rates.* Washington, DC: Education Trust.

Chang, M. J. (2002). The impact of undergraduate diversity course requirement on students' racial views and attitudes. *Journal of General Education, 51*(1), 21–42.

Chen, X., & Carroll, C. D. (2005). *First-generation students in post-secondary education: A look at their college transcripts.* Washington, DC: National Center for Education Statistics.

Chickering, A. W., & Reisser, L. (1993). *Education and Identity* (2nd ed.). San Francisco: Jossey-Bass.

Chizhik, E. W., & Chizhik, A, W. (2002). A path of social change: Examining students' responsibility, opportunity, and emotion toward social justice. *Education and Urban Society, 34*(3), 283–297.

Choy, J. (2002). *Students whose parents did not go to college: Postsecondary, access, persistence, and attainment.* Washington, DC: U.S. Department of Education, National Center for Educational Statistics.

Choy, S. P. (2000). Low-income students: Who they are and how they pay for their education. *Education Statistics Quarterly, 2*(2), 85–87.

Claxton, C. (1991). Teaching, learning and community: An interview with Parker Palmer. *Journal of Developmental Education, 15*(2), 22–25, 33.

Clinchy, B. V. (2000). Toward a more connected vision of higher education. In M. B. Magolda (Ed.), *New Directions for Teaching and Learning* (Vol. 2000, Issue 82, pp. 27–35). San Francisco: Jossey-Bass.

Conklin, M. E., & Dailey, A. R. (1981). Does consistency of parental educational encouragement matter for secondary students? *Sociology of Education, 54,* 254–262.

Constantine, M. G., & Chen, E. C. (1997). Intake concerns of racial and ethnic minority students at a university counseling center: Implications for developmental programming and outreach. *Journal of Multicultural Counseling & Development, 25*(3), 210–218.

Cook, B., & King, J. (2007). *2007 Status Report on the Pell Grant Program.* Washington, DC: American Council on Education.

Cross, K. P. (1998). Why learning communities? Why now? *About Campus, 3*(3), 4–11.

Darder, A. (1991). *Culture and power in the classroom: A critical foundation for bicultural education.* New York: Bergin and Garvey.

Darder, A. (1995). Buscando America: The contributions of critical Latino educators to the academic development and empowerment of Latino students in the U.S. In P. L. McLaren & C. E. Sleeter (Eds.), *Multicultural education, critical pedagogy, and the politics of difference* (pp. 319–348). Albany: State University of New York Press.

Darder, A. (1997). Creating the conditions for cultural democracy in the classroom. In A. Darder, R. D. Torres, & H. Gutìerrez (Eds.). *Latinos and education: A critical reader* (pp. 331–350). New York: Routledge.

Denzin, N. K., & Lincoln, Y. S. (1998). *Strategies of qualitative inquiry.* Thousand Oaks, CA: Sage.

Domizi, D. (2008). Student perceptions about their informal learning experiences in a first-year residential learning community. *Journal of the First-Year Experience & Students in Transition, 20*(1), 99–110.

Duranczyk, I. M., & Lee, A. (2007). Process pedagogy, engaging mathematics. *Academic Exchange Quarterly, 11*(2), 31–35.

Engle, J., & Tinto, V. (2008). *Moving beyond access: College for low-income, first-generation students.* Washington, DC: The Pell Institute.

Engstrom, C., & Tinto, V. (2007). *Pathway to student success: The impact of learning communities on the success of academically under-prepared college students.* Syracuse, NY: Syracuse University, prepared for the William and Flora Hewlett Foundation.

Engstrom, C., & Tinto, V. (2008). Learning better together: The impact of learning communities on the persistence of low-income students. *Opportunity Matters: A Journal of Research Informing Educational Opportunity Practice and Programs, 1*(1), 5–21.

Federal Poverty Income Guidelines (2009). *Federal Register*, 74(4), 4199–4201. Retrieved from http://www.coverageforall.org/pdf/FHCE_FedPovertyLevel.pdf

Flick, U. (1998). *An introduction to qualitative research.* London, UK: Sage.

Fogarty, J., & Dunlap, L. (with Dolan, E., Hesse, M., Mason, M., & Mott, J. (2003). *Learning communities in community colleges.* In the *National Learning Communities Project monograph series.* Olympia, WA: Washington Center for Improving the Quality of Undergraduate Education.

Freire, P. (1971). *Pedagogy of the oppressed.* New York: Continuum.

Frye, M. (2001). Oppression. In P. Rothenberg (Ed.), *Race, class, and gender in the United States* (pp. 139–163). New York: Worth Publishers.

Fuligni, A. J., Tseng, V., & Lam, M. (1999). Attitudes toward family obligations among American adolescents with Asian, Latin American, and European backgrounds. *Child Development, 70*(4), 1030–1044.

Gabelnick, F., MacGregor, J., Matthews, R. S., & Smith, B. L. (1999). *Learning Communities: Creating connection among students, faculty, and disciplines.* San Francisco: Jossey-Bass.

Gaff, J., & Lambert, L. (1996). Socializing future faculty to the values of undergraduate education. *Change, 2*(3), 38–45.

Gardner, H. (1983). *Frames of mind: The theory of multiple intelligences.* New York: Basic Books.

Gardner, H. (1998). A multiplicity of intelligences. *Scientific American Presents,* 18–23.

Gay, G. (1995). Mirror images on common issues: Parallel between multicultural education and critical pedagogy. In P. L. McLaren & C. E. Sleeter (Ed.), *Multicultural education, critical pedagogy, and the politics of difference* (pp. 155–190). New York: State University of New York Press.

Gerald, D., & Haycock, K. (2006). *Engines of inequality: Diminishing equity in the nation's premier public universities.* Washington, DC: Education Trust.

Ginorio, A., & Huston, M. (2001). *Si, se puede! Yes, we can: Latinas in school.* Washington, DC: American Association of University Women Educational Foundation.

Giroux, H. A. (1981). *Ideology, culture, and the process of schooling.* Philadelphia: Temple University Press.

Giroux, H. A., & McLaren, P. L. (1989). Introduction: Schooling cultural politics, and struggles for democracy. In H. A. Giroux & P. L. McLaren (Eds.), *Critical pedagogy, the state, and cultural struggle* (pp. 11–35). Albany: State University of New York Press.

Giroux, H. A., & Simon, R. (1989). Popular culture and critical pedagogy: Everyday life as a basis for curriculum knowledge. In H. A. Giroux & P. L. McLaren (Eds.), *Critical pedagogy, the state, and cultural struggle* (pp. 236–252). Albany: State University of New York Press.

Giroux, H. A. (1992). *Border crossings: Cultural workers and the politics of education.* New York: Routledge

Goodsell-Love, A. (1999). What are learning communities? In J. Levine (Ed.), *Learning communities: New structures, new partnerships for learning.* In the first year experience monograph series (p. 26). Columbia: University of South Carolina, National Resource Center for the First-Year Experience and Students in Transition.

Goodsell-Love, A. G., & Tokuno, K. A. (1999). Learning community models. In J. H. Levine (Ed.), *Learning communities: New structures, new partnerships for learning* (Monograph No. 26, pp. 9–18). Columbia: University of South Carolina, National Resource Center for the First-Year Experience and Students in Transition.

Grant, C. A., & Tate, W. F. (2001). Multicultural education through the lens of multicultural education research literature. In J. A. Banks, & C. A. McGee Banks (Eds.), *Handbook of research on multicultural education* (pp. 145–162). San Francisco: Jossey-Bass.

Groutt, J. (2003, January). Milestones of TRIO: Milestones of TRIO history, Part I. *Opportunity Outlook*, 21–27.

Harklau, L., & Siegal, M. (2009). Immigrant youth and higher education. In M. Roberge, M. Siegal, & L. Harklau (Eds.), *Generation 1.5 meets college composition: Issues in the teaching of writing to U.S. educated learners of ESL* (pp. 25–34). New York: Routledge.

Harklau, L., Losey, K. M., & Siegal, M. (1999). *Generation 1.5 meets college composition: Issues in the teaching of writing to U. S. educated learners of ESL.* Mahwah, NJ: Lawrence Erlbaum Associates.

Henscheid, J. (2004). *Integrating the first-year experience: The role of first-year seminars in learning communities.* Columbia: University of South Carolina, National Resource Center for the First-Year Experience and Students in Transition.

Higbee, J. L., & Lundell, D. B. (2005). An introduction to the General College. In D. B. Lundell, J. L. Higbee, & D. R. Arendale (Eds.), *The General College vision: Integrating, intellectual growth, multicultural perspectives, and student development* (pp. 7–16). Minneapolis: University of Minnesota, General College, Center for Research on Developmental Education and Urban Literacy.

Hogue, C., Parker, K., & Miller, M. (1998). Talking the walk: Ethnical pedagogy in the multicultural classroom. *Feminist Teacher, 12*(2), 89–106.

hooks, b. (1993). Transformative pedagogy and multiculturalism. In T. Perry & J. W. Fraser (Eds.), *Freedom's plow: Teaching in the multicultural classroom.* New York: Routledge.

hooks, b. (1994). *Teaching to transgress: Education as the practice of freedom.* New York: Routledge.

Horn, L., & Nunez, A.-M. (2000). Mapping the road to college: First-generation students math track, planning strategies, and context of support. *Education Statistics Quarterly, 2*(1), 81–86.

Hornak, A., & Ortiz, A. M. (2004). Creating a context to promote diversity education and self-authorship among community college students. In M. B. Baxter Magolda & P. M. King (Eds.), *Learning partnerships: Theory and models of practice to educate for self-authorship* (pp. 91–124). Sterling, VA: Stylus.

Hossler, D., & Gallagher, K. S. (1987, Spring). Studying student college choices: A three-phase model and the implications for policymakers. *College and University*, 207–221.

Howard, A. (2001). Students from poverty: Helping them making through college. *About Campus, 6*(5), 5–12.

Hunter, M. S., Tobolowsky, B. F., Gardner, J. N., Evenbeck, S. E., Pattengale, J. A., Schaller, M., & Schreiner, L. A. (2009). *Helping sophomores succeed: Understanding and improving the second-year experience.* Columbia: University of South Carolina, National Resource Center for the First-Year Experience and Students in Transition.

Hurtado, S., Inkelas, K. K., Briggs, C., & Rhee, B. S. (1997). Differences in college access and choice among racial/ethnic groups: Identifying continuing barriers. *Research in Higher Education, 38*(1), 43–75.

Hurtado, S., & Navia, C. (1997). Reconciling colleges access and the affirmative action debate. In M. Garcia (Ed.), *Affirmative action's testament of hope: Strategies for a new era in higher education* (pp. 105–129). Albany: State University of New York Press.

Inkelas, K. K., Daver, Z. E., Vogt, K. E., & Leonard, J. B. (2007). Living-learning programs and first-generation college students' academic and social transition to college. *Research in Higher Education, 48*(4), 403–433.

Ishitani, T. T. (2003). A longitudinal approach to assessing attrition behavior among first-generation students: Time-varying effects of pre-college characteristics. *Research in Higher Education, 44*(4), 433–449.

Jaffee, D. (2007). Peer cohorts and the unanticipated consequences of freshman learning communities. *College Teaching, 55*(2), 65–71.

Jaffee, D., Carle, A. C., Phillips, R., & Paltoo, L. (2008). Intended and unintended consequences of first-year learning communities: An initial investigation. *Journal of the First-Year Experience & Students in Transition, 20*(1), 53–70.

Jalomo, R. E., & Rendón, L. I. (2004). Moving to a new culture: The upside and downside of the transition to college. In. L. I. Rendón, M. Garcia, & D. Person (Eds.), *Transforming the first-year experience for students of color* (Monograph No. 38, pp. 37–54). Columbia: University of South Carolina, National Resource Center for the First-Year and Students in Transition.

James, P. (2000). "I am the dark forest": Personal analogy as a way to understand metaphor. *Art Education, 53*(5), 6–11.

James, P. (2002). *1481 Creative Art Lab: Masked Transformation*. Spring class syllabus, University of Minnesota, General College, Minneapolis.

James, P. (2005). Aesthetic, metaphoric, creative, and critical thinking: The arts in the General College. In D. B. Lundell, J. L. Higbee, & D. R. Arendale (Eds.), *The General College vision: Integrating intellectual growth, multicultural perspectives, and student development* (pp. 247–285). Minneapolis: University of Minnesota, General College, Center for Research on Developmental Education and Urban Literacy.

James, P. A., Bruch, P. L., & Jehangir, R. R. (2006). Ideas in practice: Building bridges in a multicultural learning community. *Journal of Development Education*, *29*(3), 10–18.

Jarmon, B., Brunson, D. A., & Lampl, L. L. (2007). The power of narratives in the process of teaching and learning about diversity. In D. A. Brunson, B. Jarmon, & L. L. Lampl (Eds.), *Letters from the future: Linking students and teaching with the diversity and everyday life* (pp. 27–45). Sterling, VA: Stylus.

Jehangir, R. (2002a). Charting a new course: Learning communities and universal design. In J. L. Higbee. (Ed.), *Curriculum transformation and disability: Implementing universal design in higher education*. Minneapolis, MN: University of Minnesota, General College, Center for Research in Developmental Education and Urban Literacy.

Jehangir, R. (2002b). Higher education for whom: The battle to include developmental education at the four-year university. In D. B. Lundell, J. L. Higbee, & I. M. Duranczyk (Eds.), *Developmental education: Policy and practice*. Warrensburg, MO: National Association of Developmental Education.

Jehangir, R. (2008). In their own words: Voices of first generation students in a multicultural learning community. *Opportunity Matters*, *1*(1), 22–31.

Jehangir, R. (2009a). Cultivating voice: First generation students seek full academic citizenship in multicultural learning communities. *Innovative Higher Education*, *34*(1), 33–49.

Jehangir, R. (2009b). In search of place: Using case studies to create counter-hegemonic spaces in higher education. In D. Cleveland (Ed.), *When "minorities are strongly encouraged to apply": Diversity and affirmative action in higher education* (Vol. 15, pp. 281–288). New York: Peter Lang.

Jehangir, R. (2010a). Stories as knowledge: Bringing the lived experience of first-generation college students into the academy. *Urban Education*, *45*(4), 533–553.

Jehangir, R. (2010b). *Learning communities an critical multicultural pedagogy: Narratives of first-generation college students in a public White university*. Manuscript submitted for publication.

Jehangir, R. (2010c). *Multicultural learning communities: Vehicles for developing self-authorship in first-generation college students*. Manuscript submitted for publication.

Jehangir, R. (Forthcoming 2012). The influence of multicultural learning communities on the intrapersonal development of first-generation college students. *Journal of College Student Development*.

Johnson, A. B. (2005). From the beginning: This history of developmental education and the pre-1932 General College idea. In D. B. Lundell, J. L. Higbee, & D. R. Arendale (Eds.), *The General College vision: Integrating, intellectual growth, multicultural perspectives, and student development* (pp. 39–60). Minneapolis: University of Minnesota, General College, Center for Research on Developmental Education and Urban Literacy.

Johnson, D. W., & Johnson, R. T. (1991). *Learning together and alone: Cooperative, competitive, and individualistic learning* (3rd ed.). Needham Heights, MA: Allyn and Bacon.

Kanpol, B., & McLaren, P. (1995). *Critical multiculturalism: Uncommon voices in a common struggle.* Westport, CT: Bergin and Garvey.

Karabel, J. (1986). Community college and social stratification in the 1980s. In L. S. Zwerling (Ed.), *The community college and its critics: New directions for community colleges* (Vol. 54, pp. 12–30). San Francisco: Jossey-Bass.

Kegan, R. (1994). *In over our heads: The mental demands of modern life.* Cambridge, MA: Harvard University.

Kellogg Commission on the Future of State and Land-Grant Universities (2001). *Returning to our roots.* Executive summaries of the reports of the Kellogg Commission on the Future of State and Land-Grant Universities. Washington, DC.: National Association of State and Land-Grant Universities.

Kelly, P. (2005). *As America becomes more diverse: The impact of state higher education inequality.* Boulder, CO: National Center for Higher Education Management Systems.

Kerr, C. (1995). *The uses of the university* (4th ed.). Cambridge, MA: Harvard University Press.

Kiang, P. N. (1992). Issues of curriculum and community for first-generation Asian Americans in college. In L. S. Zweling & H. B. London (Eds.), *First-generation students: Confronting the cultural issues* (Vol. 80, pp. 97–112). San Francisco: Jossey-Bass.

Kincheloe, J. L., & McLaren, P. (2000). Rethinking critical theory and qualitative research. In N. K. Denzin & Y. S. Lincoln (Eds.), *The handbook of qualitative research* (2nd ed., pp. 279–313). Thousand Oaks, CA: Sage.

King, P. M., & Shuford, B. C. (1996). A multicultural view is a more cognitively complex view: Cognitive development and multicultural education. *American Behavior Scientists, 40*(2), 153–164.

Kinzie, J., & Kuh, G. D. (2004). Going deep: Learning from campuses that share responsibility for student success. *About Campus, 9*(5), 2–8.

Kuh, G. D. (2003). What we're learning about student engagement from NSSE. *Change, 35*(2), 24–32.

Kuh, G. D. (2008). *High-impact education practices.* Washington, DC: Association of American Colleges and Universities.

Kuh, G. D., Kinzie, J., Schuh, J., & Whitt, E. J. (2005). *Student success in college: Creating conditions that matter.* Washington, DC: Jossey-Bass.

Kuh, G. D., & Whitt, E. J. (1988). *The invisible tapestry: Culture in American colleges and universities.* ASHE-ERIC higher education report (Vol. 17). Washington, DC: George Washington University, Graduate School of Education.

Lara, J. (1992). Reflections: Bridging cultures. In L. S. Zwerling & H. B. London (Eds.), *First generation college students: Confronting the cultural issues.* New Directions for Teaching and Learning, No. 80. San Francisco: Jossey-Bass

Lardner, E. D. (2003). Approaching diversity through learning communities. *Washington Center occasional paper for improving the quality of undergraduate education,* (2). Retrieved from http://www.evergreen.edu/washcenter/publications.asp

Lardner, E. D. (2005). *Diversity, educational equity, and learning communities.* Learning communities and educational reform series. Olympia, WA: Washington Center for Improving the Quality of Undergraduate Education.

Lardner, E. D., & Malnarich, G. (2008). A new era in learning: Why the pedagogy of intentional integration matters. *Change, 40*(4), 30–37.

Laufgraben, J. L. (2003). Faculty development: Growing, reflecting, learning, and changing. In J. O'Connor (Ed.), *Learning communities in research universities.* In The National Learning Communities Project monograph series (pp. 33–38). Olympia, WA: Washington Center for Improving the Quality of Undergraduate Education, in cooperation with the American Association for Higher Education.

Law, C. L. (1995). Introduction. In C. L. Law & C. L. B. Dews (Eds.), *This fine place so far from home: Voices of academics from the working class.* Philadelphia: Temple University Press.

Lenning, O. T., & Ebbers, L. H. (1999). The powerful potential of learning communities: Improving education for the future. *ASHE-ERIC Higher Education Report* (Vol. 26). Washington, DC: George Washington University, Graduate School of Education and Human Development.

Levin, J. (2007). *Non-traditional students and community colleges: The conflict of justice and neoliberalism.* New York: Palgrave Macmillan.

Levine, J., Smith, B. L., Tinto, V., & Gardner, J. (1999). *Learning about learning communities: Taking student learning seriously.* Teleconference resource packet. Columbia, SC: University of South Carolina, National Resource Center for the First-Year Experience and Students in Transition.

Levy-Warren, M. H. (1988). Moving to new culture: Cultural identity, loss, and mourning. In J. Bloom & S. B. Bloom-Feschbach (Eds.), *The psychology of separation and loss.* San Francisco: Jossey-Bass.

Lincoln, Y. S., & Guba, E. G. (1985). *Naturalistic inquiry.* Thousand Oaks, CA: Sage.

Lohfink, M., & Paulsen, M. B. (2005). Comparing the determinants of persistence for first-generation and continuing-generation students. *Journal of College Student Development, 46*(4), 409–428.

London, H. B. (1989). Breaking away: A study of first-generation college students and their families. *American Journal of Education, 97*(1), 144–170.

London, H. B. (1992). Cultural challenges faced by first generation college students. In L. S. Zwerling & H. B. London (Eds.), *First generation college students: Confronting the cultural issues.* San Francisco, CA: Jossey-Bass.

London, H. B. (1996). How college affects first-generation college students. *About Campus, 1*(5), 9–13.

Longwell-Grice, R., & Longwell-Grice, H. (2007). Testing Tinto: How do retention theories work for first-generation, working-class students? *Journal of College Student Retention: Research, Theory, & Practice*, *9*(4), 407–420.
Magolda, P. M. (2001). What our rituals tell us about community on campus: A look at the campus tour. *About Campus*, *5*(6), 2–8.
Martorell, P., & McFarlin, I. (2007). *Help or hindrance? The effects of college remediation on academic and labor market outcomes.* Unpublished manuscript.
McDonough, P. M. (1997). *Choosing colleges: How social class and schools structure opportunity.* Albany: State University of New York Press.
McLaren, P. (1998). The pedagogy of Che Guevara: Critical pedagogy and globalization thirty years after Che. *Cultural Circles*, *3*, 29–103.
McPherson, M. S., & Schapiro, M. O. (1998). *The student aid game: Meeting need and rewarding talent in American higher education.* Princeton, NJ: Princeton University Press.
Mellow, G., van Slyck, P., & Eynon, G. (2003). The face of the future: Engaging in diversity at LaGuardia Community College. *Change*, *35*(2), 10–17.
Merriam, S. B. A. (1998). *Qualitative research and case-study applications in education.* San Francisco: Jossey-Bass.
Merriam, S. B. A. (2002). *Qualitative research in practice: Examples for discussion and analysis.* San Francisco: Jossey-Bass.
Mezirow, J. (1997). Transformation theory out of context. *Adult Education Quarterly*, *48*(1), 60–62.
Miksch, K. L., Bruch, P. L., Higbee, J. L., Jehangir, R., & Lundell, D. B. (2003). The centrality of multiculturalism in developmental education. In J. L. Higbee, D. B. Lundell, & I. M. Duranczyk (Eds.), *Multiculturalism in developmental education.* Minneapolis: University of Minnesota, General College, Center for Research on Developmental Education and Urban Literacy.
Miles, M. B., & Huberman, A. M. (1994). *Qualitative data analysis.* Thousand Oaks, CA: Sage.
Moll, L. C., Amanti, C., Neff, D., & Gonzales, N. (1992). Funds of knowledge for teaching: Using a qualitative approach to connect homes and classrooms. *Theory into Practice*, *31*(2), 132–141.
Mortenson, T. G. (2003). Economic segregation of higher education opportunity, 1973 to 2001. In *Postsecondary education opportunity* (Vol. 136). Oskaloosa, IA: Pell Institute.
Mortenson, T. G. (2006). Unmet financial need of undergraduate students by state, sector, status and income levels 2003–04. *Postsecondary education opportunity* (Vol. 163). Oskaloosa, IA: Pell Institute.
Mortenson, T. G. (2008). Family income and educational attainment 1970 to 2007. *Postsecondary education opportunity* (Vol. 197). Oskaloosa, IA: Pell Institute.
Murphy, P. E. (1981). Consumer buying roles on college choice: Parents' and students' perceptions. *College and University*, *56*(2), 140–150.
Nash, R. J. (2009). Crossover pedagogy: The collaborative search for meaning. *About Campus*, *14*(1), 2–9.

National Center for Education Statistics (NCES). (2007). *The condition of education 2007.* Washington, DC: Author.

National Center for Education Statistics (NCES). (2007a). *Digest of Educational Statistics 2006.* Washington, DC: U.S. Department of Education.

Nora, A., & Cabrera, A. F. (1996). The role of perceptions of prejudice and discrimination on the adjustment of minority students to college. *Journal of Higher Education, 67*(2), 119–148.

Nunez, A., & Cuccaro-Alamin, S. (1998). First-generation students: Undergraduates whose parents never enrolled in postsecondary education. Washington, DC: National Center for Education Statistics.

O'Connor, J. (2003). Learning communities in research universities. *The National Learning Communities Project monograph series.* Olympia, WA: Washington Center for Improving the Quality of Undergraduate Education in cooperation with the American Association for Higher Education.

Ogbu, J. U. (2001). Understanding cultural diversity & learning. In J. A. Banks & C. M. Banks (Eds.), *Handbook on research on multicultural education.* San Francisco: Jossey-Bass.

Omni, M., & Winant, H. (1994). *Racial formation in the United States from the 1960s to the 1990s.* New York: Routledge.

Omni, M., & Winant, H. (2001). Racial formations. In P. S. Rothenberg (Ed.), *Race, Class, and Gender in the United States* (5th ed., pp. 11–20). New York: Worth Publishers.

Ontai, L. L., & Raffaelli, M. (2004). Gender socialization in Latino/a families: Results from two retrospective studies. *Sex Roles: A Journal of Research, 50*(5/6), 287–299.

Ortiz, A. M. (1995). *Promoting racial understanding in college students: A study of educational and development interventions.* Paper presented at the annual meeting of the Association for the Study of Higher Education, Orlando, FL.

Ortiz, A. M. (2000). Expressing cultural identity in the learning community: Opportunities and challenges. In M. B. Magolda (Ed.), *New directions for teaching and learning* (Vol. 82, pp. 67–79). San Francisco: Jossey-Bass.

Osei-Kofi, N., Richards, S. L., & Smith, D. G. (2004). Inclusion, reflection, and the politics of knowledge: On working toward the realization of inclusive classroom environments. In. L. I. Rendón, M. Garcia, & D. Person (Eds.), *Transforming the first year of college for students of color* (pp. 55–66). Columbia: University of South Carolina, National Resource Center for the First-Year Experience and Students in Transition.

Padron, E. J. (1992). The challenge of first-generation college students: A Miami-Dade perspective. In L. S. Zwerling & H. B. London (Eds.), *First-generation students: Confronting the cultural issues* (Vol. 80, pp. 71–80). San Francisco: Jossey-Bass.

Palmer, P. (1998). *The courage to teach: Exploring the inner landscape of a teacher's life.* San Francisco: Jossey-Bass.

Parks, S. D. (2000). *Big questions, worthy dreams: Mentoring young adults in their search for meaning, purpose, and faith.* San Francisco: Jossey-Bass.

Pascarella, E. T., Wolniak, G. C., Pierson, C. T., & Terenzini, P. T. (2003). Experiences and outcomes of first-generation students in community colleges. *Journal of College Student Development, 44*(3), 420–429.

Pascarella, E. T., Pierson, C. T., Wolniak, G. C., & Terenzini, P. T. (2004). First-generation college students: Additional evidence on college experiences and outcomes. *Journal of Higher Education, 75*(3), 249–284.

Pérez, P. A., & McDonough, P. M. (2008). Understanding Latina and Latino college choice. *Journal of Hispanic Higher Education, 7*(3), 249–265. doi: 10.1177/1538192708317620

Perry, W. G. (1999). *Forms of ethnical and intellectual development in the college years: A scheme.* San Francisco: Jossey-Bass.

Philippe, K. A., & Valiga, M. J. (2000). Faces of the future: A portrait of America's community college students. *Community College Review, 32*(1), 1–20.

Phinney, J. S., Ong, A., & Madden, T. (2000). Cultural values and intergenerational value discrepancies in immigrant and non-immigrant families. *Child Development, 71*, 528–539.

Pizzolato, J. E. (2003). Developing self-authorship: Exploring the experiences of high-risk college students. *Journal of College Student Development, 44*(6) 797–812.

Pizzolato, J. E. (2004). Coping with conflict: Self-authorship, coping, and adaptation to college in first-year, high-risks students. *Journal of College Student Development, 45*, 425–442.

Rendón, L. I. (1992). From the barrio to the academy: Revelations of a Mexican American "scholarship girl." In. L. S. Zwerling & H. B London (Eds.), *First generation college students: Confronting the cultural issues* (pp. 55–64). San Francisco: Jossey-Bass.

Rendón, L. I. (1994). Validating culturally diverse students: Toward a new model of learning and student development. *Innovative Higher Education, 19*(1), 33–51.

Rendón, L. I. (1996). Life on the border. *About Campus, 1*(5), 14–20.

Rendón, L. I. (1998). Helping nontraditional students be successful in college. *About Campus, 3*(1), 2–3.

Rendón, L. I. (2009). *Sentipensante (sensing/thinking) pedagogy: Education for wholeness, social justice, and liberation.* Sterling, VA: Stylus.

Renner, K. E. (2003). Racial equity and higher education. *Academe, 89*(1), 38–43.

Rice, R. E. (1996). *Making a place for the new American scholar.* Washington, DC: American Association of Higher Education.

Richardson, F. C. (1994). The president's role in the shaping the culture of academic institutions. In J. D. Davis (Ed.), *Coloring the walls of ivy: Leadership and diversity in the academy* (pp. 14–35). Boston: Anker Publishing Company.

Richardson, R. C., & Skinner, E. F. (1992). Helping first-generation minority students achieve degrees. In L. S. Zwerling & H. B. London (Ed.), *First generation college students: Confronting the cultural issues* (pp. 29–43). San Francisco: Jossey-Bass.

Rivera, J., & Poplin, M. (1995). Multicultural, critical, feminine, and constructive pedagogies seen through the lives of youth: A call for the revisioning of these and beyond: Toward a pedagogy for the next century. In P. L. McLaren & C. E. Sleeter (Ed.), *Multicultural education, critical pedagogy, and the politics of difference* (pp. 221–244). Albany: State University of New York Press.

Roberge, M., Siegal, M., &Harklau, L.(Eds.)(2009) *Generation 1.5 meets college composition: Issues in the teaching of writing to U.S. educated learners of ESL* New York: Routledge.

Rodriguez, R. (1982). *Hunger of memory: The education of Richard Rodriguez: An autobiography.* New York: Bantam Books.

Rose, M. (2009). The reciprocal nature of education. *About Campus, 14*(5), 1–32.

Ross, K. (1993). The world literature and cultural studies program. *Critical inquiry, 19*(4), 666–676.

Roueche, J. E., & Roueche, S. D. (1999). *High stakes, high performance: Making remedial education work.* Washington, DC: Community College Press.

Ruppert, S. R. (2003). *Closing the college participation gaps: A national summary.* Denver, CO: Education Commission of the States.

Sacks, P. (2007). *Tearing down the gates: Confronting the class divide in American education.* Berkeley: University of California Press.

Saldana, D. H. (1994). Acculturative stress: Minority status and distress. *Hispanic Journal of Behavioral Sciences, 16*(2), 117–124.

Schelske, B & Schelske, S. (2002). *TRIO Student Support Services Grant 1998–2002.* University of Minnesota, General College. Minneapolis: University of Minnesota.

Schmader, T., Major, B., & Gramzow, R. W. (2001). Coping with ethnic stereotype in the academic domain: Perceived injustice and psychological disengagement. *Journal of Social Issues, 57*(1), 93–111.

Schoem, D. (2003). Learning communities at the University of Michigan: The best of both worlds. In J. O'Connor (Eds.), *Learning communities in research universities. The National Learning Communities Project monograph series* (pp. 43–46). Olympia, WA: Washington Center for Improving the Quality of Undergraduate Education, in cooperation with the American Association for Higher Education.

Schoem, D., & Pasque, P. A. (2003). Learning for the common good: A diverse learning community lives and learns together in the Michigan community scholars program. *About Campus, 8*(1), 9–16.

Schön, D. A. (1991). *The reflective turn: Case studies in and on educational practice.* New York: Columbia University, Teachers College Press.

Schroeder, C. C., Minor, F. D., & Tarkow, T. A. (1999). Learning communities: Partnerships between academic and student affairs. In J. Levine (Ed.), *Learning communities: New structures, new partnerships for learning* (Vol. 26). Columbia: University of South Carolina Press.

Scrivener, S., Bloom, D., LeBlanc, A., Paxson, C., Rouse, C. E., & Sommo, C. (with Au, J., Teres, J. J., & Yeh, S.). (2008). *A good start: Two-year effects of a freshman learning community program at Kingsborough Community College.* New York: MDRC.

Shapiro, N. S. (2003). University of Maryland College Park scholars: Creating a coherent lens for general education. In J. O'Connor (Ed.), *Learning Communities in Research Universities.* In *The National Learning Communities Project monograph series* (pp. 39–42). Olympia, WA: Washington Center for Improving the Quality of Undergraduate Education, in cooperation with the American Association for Higher Education.

Shapiro, N. S., & Levine, J. H. (1999). *Creating learning communities: A practical guide to winning support, organizing change, and implementing programs.* San Francisco: Jossey-Bass.

Shor, I. (1993). Education is politics: Paulo Freire's critical pedagogy. In P. McLaren & P. Leonard (Eds.), *Paulo Freire: A critical encounter.* New York: Routledge.

Sleeter, C. E. (1995). Refections on my use of multicultural and critical pedagogy when students are White. In P. L. McLaren & C. E. Sleeter (Eds.), *Multicultural education, critical pedagogy, and the politics of difference* (pp. 415–434). New York: State University of New York Press.

Sleeter, C. E., & Grant, C. (1988). *Making choices for multicultural education: Five approaches to race, class and gender.* Columbus, OH: Merill.

Sleeter, C. E., & McLaren, P. L. (1995). Introduction: Exploring connection to build a critical multiculturalism. In P. L. McLaren & C. E. Sleeter (Eds.), *Multicultural education, critical pedagogy, and the politics of difference* (pp. 5–28). Albany: State University of New York Press.

Smith, B. L. (1991). Taking structure seriously: The learning community model. *Liberal Education, 77*(2), 42–48.

Smith, B. L., MacGregor, J., Matthews, R., & Gabelnick, F. (2004). *Learning communities: Reforming undergraduate education.* San Francisco: Jossey-Bass.

Smoke, T., & Haas, T. (1995). Ideas in practice: Linking classes to develop students' academic voices. *Journal of Developmental Education, 19*(2), 28–32.

Solomon, Y. (2009). *Mathematical literacy: Developing identities of inclusion.* New York: Routledge.

St. John, E. (2002). *The access challenge: Rethinking the causes of the new inequality.* Bloomington, IN: Indiana Education Policy Institute.

St. John, E. (2005). *Affordability of post-secondary education: Equity and adequacy across the 50 states.* Washington, DC: The Center for American Progress.

Staats, S. K. (2005). Multicultural mathematics: A social issues perspective in lesson planning. In D. B. Lundell, J. I. Higbee, & D. R. Arendale (Eds.), *The General College vision: Integrating intellectual growth, multicultural perspectives, and student development* (pp. 185–199). Minneapolis: University of Minnesota, General College, Center for Research on Developmental Education and Urban Literacy.

Stage, F. K., & Hossler, D. (1989). Differences in family influences on college attendance plans for male and female ninth graders. *Research in Higher Education, 30*(3), 301–315.

Stake, R. E. (1995). *The art of case-study research.* Thousand Oaks, CA: Sage.

Stefanou, C. R., & Salisbury-Glennon, J. D. (2002). Developing motivation and cognitive learning strategies through an undergraduate learning community. *Learning Environments Research, 5*, 77–79.

Sy, S. R., & Romero, J. (2008). Family responsibilities among Latina college students from immigrant families. *Journal of Hispanic Higher Education, 7*(3), 212–227.

Takaki, R. (1993). *A different mirror: A history of multicultural America.* New York: Little Brown and Company.

Terenzini, P. T., Cabrera, A. F., & Bernal, E. M. (2001). *Swimming against the tide: The poor in American higher education.* New York: College Board.

Terenzini, P. T., Rendón, L. I., Upcraft, M. L., Millar, S. B., Allison, K. A., Gregg, P. L., & Jalomo, R. (1994). The transition to college: Diverse students, diverse stories. *Research in Higher Education, 35*(1), 57–73.

Terenzini, P. T., Springer, L., Yaeger, P. M., Pascarella, E. T. & Nora, A. 1996). First-generation college students: Characteristics, experiences, and cognitive development. *Research in Higher Education, 37*(1), 1–22.

Tierney, W. G. (1997). The parameters of affirmative action: Equity and excellence in the academy. *Review of Educational Research, 67*(2), 165–196.

Tinto, V. (1987). *Leaving college: Rethinking the causes and cure of student attrition.* Chicago, IL: University of Chicago Press.

Tinto, V. (1997). Classrooms as communities: Exploring the educational character of student persistence. *Journal of Higher Education, 68*(6), 599–623.

Tinto, V. (1998a). Colleges as communities: Taking research on student persistence seriously. *The Review of Higher Education, 21*(2), 167–177.

Tinto, V. (1998b. *Learning communities and the re-construction of remedial education in higher education.* Paper presented at the Conference on Replacing Remediation in Higher Education, Stanford University, Stanford, CA.

Tinto, V., Goodsell-Love, A., & Russo, P. (1993). Building community. *Liberal Education, 79*(4), 16–21.

Tinto, V., Russo, P., & Kadel, S. (1994). Constructing educational communities: Increasing retention in challenging circumstances. *Community College Journal, 64*, 26–30.

Tobin, E. M. (2009). The modern evolution of America's flagship universities. In W. G. Bowen, M. M. Chingos, & M. S. McPherson (Eds.), *Crossing the finish line* (pp. 239–264). Princeton, NJ: Princeton University Press.

Tompkins, J. (1990). Pedagogy of the distressed. *College English 52*(6), 653–660.

Torres, V., & Hernandez, E. (2007). The influence of ethnic identity development on self-authorship: A longitudinal study of Latino/a college students. *Journal of College Student Development, 48*(5), 558–573.

Tussman, J. (1969). *Experiment at Berkeley.* London, UK: Oxford University Press.

University of Minnesota. (2004). General College fact sheet. Retrieved March 3, 2004, from http://www.gen.umn.edu/gc/images/fact_sheet.pdf

University of Minnesota, General College. (1997). *External review report.* Minneapolis: University of Minnesota.

U.S. Census Bureau (2003). *Poverty in the United States: 2002.* Washington, DC: U.S. Government Printing Office.

U.S. Department of Education (2006). A test of leadership: Charting the future of higher education. A report of the Commission appointed by Secretary of Education Margaret Spellings. Retrieved online at http://www 2.ed.gov/about/bdscomm/list/hiedfuture/index.html

U.S. Department of Education (2010b). Federal TRIO programs current year income level.

http://www.ed.gov/about/offices/list/ope/trio/incomelevels.html . Retrieved on Jan 9th 2010

U.S. Department of Education (2010a): Federal TRIO programs Home page http://www.ed.gov/about/offices/list/ope/trio/index.html. Retrieved on Jan 9th 2010

Van Maanen, M. (1990). *Researching lived experience: Human science for an action sensitive pedagogy.* London, UK: Althouse Press.

Vander Putten, J. (2001). Bringing social class to the diversity challenge. *About Campus*, *6*(5), 14–19.

Villaverde, L. E. (2007). Dialogue, reflection, and critical analysis. In B. Jarmon, D. A. Brunson, & L. L. Lampl (Eds.), *Letters from the future: Linking students and teaching with the diversity of everyday life* (pp. 206–248). Sterling, VA: Stylus.

Visher, M. G., Wathington, H., Richburg-Hayes, L., & Schneider, E. (2008). The learning communities demonstration rationale, sites, and research design. National Center for Postsecondary Research. Teachers College, Columbia University, NY 10027. Retrieved from ERIC, http://eric.ed.gov/ERICWeb Portal/search/detailmini.jsp?_nfpb=true&_&ERICExtSearch_SearchValu e_0=ED501563&ERICExtSearch_SearchType_0=no&accno=ED501563

Wallerstein, N. (1987). Problem-posing education: Freire's method for transformation. In I. Shor (Ed.), *Freire for the classroom* (pp. 33–44). Portsmouth, NH: Boynton/Cook.

Warburton, E., Bugarin, R., & Nunez, A. M. (2001). *Bridging the gap: Academic preparation and post secondary success of first-generation college students.* Washington, DC: National Center for Educational Statistics.

Weber, L. (1968). A conceptual framework for understanding race, class, gender, and sexuality. *Psychology of Women Quarterly*, *22*, 13–22.

Williams, D. A., Berger, J. B., & McClendon, S. A. (2005). *Toward a model of inclusive excellence and change in post-secondary institutions.* Washington, DC: Association of American Colleges and Universities.

Wilson, J. E. (1984). Balancing class locations. In J. Ryan & C. S. Sackrey (Eds.), *Strangers in paradise: Academics from the working class* (pp. 207–218). Boston: South End Press.

Wink, J. (2000). *Critical pedagogy.* New York: Addison Wesley Longman.

Wirt, J., Choy, S., Rooney, P., Provasnik, S., Sen, A., & Tobin, R. (2004). *The condition of education 2004.* Jessup, MD: National Center for Educational Statistics, U.S. Department of Education.

Wright, F. L. (n.d.). [Quotation.] Retrieved from http://www.brainyquote.com/quotes/quotes/f/franklloyd143144.html

York-Anderson, D. C., & Bowman, S. L. (1991). Assessing the college knowledge of first-generation and second-generation college students. *Journal of College Student Development, 32*(2), 116–122.

Yosso, T. J. (2005). Whose culture has capital? A critical race theory discussion of community cultural wealth. *Race, Ethnicity, and Education, 8*(1), 69–91.

Zea, M. C., Reisen, C. A., Beil, C., & Caplan, R. D. (1997). Predicting intention to remain in college among ethnic minority and non-minority students. *Journal of Social Psychology, 137*(2), 149–160.

Zhao, C., & Kuh, G. D. (2004). Adding value: Learning communities and student engagement. *Research in Higher Education, 45*(2), 115–138.

Index

academia, strangers to, 132
academic
  dialogue, 79, 126
  identity, 9, 70, 184
  preparation, 4, 81
  voice, 139–40, 159–60
academic language. *See* language: academic
acculturation, forced, 141
active teaching and learning, 180
adaptation. *See* transition
adjustment. *See* transition
admissions, 18, 49,
  raising criteria of, 38
  structural gate, 38
adult learners, 48
advisors, 89–90, 96–97, 126–29, 171, 181
African Americans, 10, 32, 39, 98
age, 53, 115
agency, 5, 54, 69, 97–98, 102–4, 106, 110
artifacts
  definition, 33
Asian Americans, 10, 98
assessment, 76
assignments, shared, 78
attrition, 22, 81
  FG vs. second-generation college students, 22

belonging, sense of, 83–84, 105, 123, 126, 132, 135–37, 151
biracial students, 10, 114
Blacks, 10, 30, 98, 114. *See also* African Americans
blue collar, 15. *See also* class: working
Bradley Learning Community, 80
bridge-building, 86
Brubacher, John, 47

campus
  full citizenship on, 178
  modus operandi, 6, 169
capital
  cultural, 32–33, 54–57, 69, 88, 103, 108, 113, 143
  social, 22
case study, interpretive multiple, 8
Chadbourne Residential College, 80
Chicana, 106, 111. *See also* Hispanic Americans
claiming the self, 9, 138
class
  classism, 16, 41, 51, 62, 65, 69, 102, 104, 135, 161
  working, 15, 174, 197, 205
classroom dynamics, 38, 40, 110, 112
coalitions, 53, 181–82
  between classrooms and campus, 182
codebook, 6, 29–43. *See also* expectations
Coffman, Lotus D., 99
cognitive dissonance, 158
cohorts. *See* peer groups
co-learners, 56, 83, 95, 103, 107, 126, 173, 177

collaboration, 74, 87, 96, 105, 113
College English Transitions. *See* Postsecondary Teaching and Learning
College of Education and Human Development, 99
comfort zone, 108, 125, 127, 141, 158, 169–70
Commanding English. *See* Postsecondary Teaching and Learning
communities
  curricular learning, 78
  inclusive, 6, 91
  interdisciplinary learning, 5
  mentoring, 182
  teaching, 181
community colleges, 47, 80–81, 84
compensatory add-ons, 54
conflict, 151
  catalyst for learning, 65
  problem posing, 61–63, 88
contemplation, 144
courses, MLC, 8
  Creativity Art Lab, 96
  International Literature, 96, 101
  Multicultural Relations, 101, 102
  Writing Lab, 102
  See also Multicultural Learning Community
courses, team-taught, 78
coursework, interrelated, 78
critiques
  of students' academic experiences, 7, 10, 115, 165, 168, 171
cultural
  brokers, 25, 185
  dislocation and relocation, 41
  role reversal, 25
curriculum
  content integration, 59
  GLBT, 155
  hidden, 61, 63
  multicultural, 7, 60, 70, 132, 136, 156, 180

debt, 17–19
  cumulative loan debt, 19
  loan default, 117, 165
deconstruction of, 98
  identities, 61
  isms, 51
  teaching, 103
design, learning-community. *See* praxis: learning-community design
developmental courses, 81–82
  overrepresentation of FG students, 20
Dewey, John, 28, 76
disability, 14, 52–53, 97
  services, 182
discrimination, 26, 41, 143, 154
  experience of, 26
disequilibrium. *See* self-authorship
diversity, 10, 53, 81, 83–85, 87, 88, 108, 115, 132–33, 152, 176, 177
  definition, 53
  dovetailing, 87

education
  experiential, 76
  as the great equalizer, 47, 168
  as a public good, 6, 43, 48
  reform, 53, 74, 91, 185
  student-centered, 76
  transformational, 53
Emerging Scholars programs, 82
engaged disagreement, 151. *See also* conflict
engaged place and space, 48
English, nonnative speakers of, 14, 35
ethnicity, 27, 57, 71, 81, 84, 109, 122, 133
expectations, 13, 21–24, 34–36, 41–42, 78–79, 128–29
  explicit and implicit, 6, 30
  low, 39
  norms, middle-class, 31, 33
  parental, 22
Experiment at Berkeley, 76

faculty
  development opportunities, 180
  perception of FG students, 39
  positionality, 173
  relationships with FG students, 38
  self-fulfilling prophecy, 39
family
  conflict, 42, 60
  loyalty to, 42
  parental educational level, 21
  renegotiating relationships with, 43
  role of parents, 21
financial aid, 18–19, 49, 90, 100,
  financial need vs. merit-based aid, 18
  loan vs. grant funding, 18
  nonsubsidized loans and tax credits, 18
first generation (FG), 14, 16, 131
  definition, 14
  demographics, 14
  dropout rates in college, 17
  enrollment in postsecondary education, 16, 17, 30
  family income, 15, 18, 22–23
  obstacles, 2, 26, 48, 120
forced acculturation. *See* acculturation, forced
framework of neutrality. *See* knowledge
Freire, Paolo, 50, 65

gender, 10, 52–53, 97–98, 132–33, 150, 154, 156
  female students, 10, 98
  sexism, 62, 88, 102, 104, 154, 161
General College, 99, 101, 170
  history of, 99
Guevara, Che, 50

hegemony, 50, 67
heterosexism, 88. *See* homophobia
higher education
  access to, 3, 47
  cultural expectations and norms, 31
  democracy and, 7, 50, 53, 61, 74, 76–77, 184
  entitlement vs. access to, 47
  goal of, 47
  philosophical differences, 48
  role of, 28, 34, 48
  types of, 47–48
high schools, 4, 20–21, 82, 113, 179
  advanced placement courses, 20
  completion rates, 15
  curricula, 20
  precollegiate experiences, 14, 19
Hispanic Americans, 10, 27, 30, 82, 98, 114
historic social memory, 6
homophobia, 62, 154. *See* heterosexism
hyperbonding, 110–11

identities
  finding congruence between multiple, 146–48, 162–63
  multiple, 4, 47, 59, 60–61, 70–71, 107–8, 140–42, 162–63, 181–82
  public vs. private, 42
immigrants, 2, 10, 32, 35, 39, 42, 48, 136
  African, 10, 39, 66, 112, 114
  Eritrea, 10
  Ethiopian, 10, 118, 137, 162
  Hmong, 10, 60, 63, 71, 137
  Liberian, 10, 117, 148
  Somalia, 10
imposter syndrome, 41
inclusive spaces. *See* engaged place and space
income status, 14–15
  family income, 15, 18, 22–23
inequity, 5, 16, 34, 52, 88
institutions
  for-profit, 16–17
  two-year, 16–18, 47, 76, 84
integration vs. separation, 183
intellectual reciprocity, 67

interaction, individualized, 135
interview research, 4, 9, 10, 41, 84, 113–15, 119, 139
  categorical aggregation, 9
  data analysis procedures, 9, 114
  emergent construction, 114
  saturation of categories, 9, 114
isolation. *See* marginalization

journey, college, 6, 129
  predisposition, 21
  preparation, 21
  matriculation, 21

Kingsborough Community College, 83
knowledge
  "banking" theory, 67, 161
  construction, 5–6, 55–56, 58, 96, 129, 140
  "legitimate," 54
  as objective, 39

LaGuardia Community College, 88
language
  academic, 34
  as gatekeeper, 34
Latina, 10, 22, 141. *See also* Hispanic Americans
learning
  collateral, 172
  cooperative, 77–78
  dialogical, 88
  interdisciplinary approaches to, 74
  logs, 8, 70, 109–10
  outcomes, 78
  problem solving, 74
  shared, 67, 70, 72, 83, 97
  styles, 66, 137, 158
  transformative, 54–55, 74, 76, 86
learning communities
  curricular learning communities, 78
  and FG students, 80–82
  history of, 76
  living-learning, 76, 78–80, 85, 90
  practice, 76, 78, 90, 91
  research on, 82, 84–85, 87
  role of, 75–76
  student-type, 79
Lewin, Kurt, 77
lived experience, 4–7, 39, 56–59, 61, 63–65, 86–88, 152–53, 158
lived history, 36
living-learning communities. *See* learning communities
low-income (LI), 15, 17, 30, 71, 81, 84, 99, 100, 132, 167

marginalization, 4–7, 39–30, 32–33, 36, 42, 48, 55, 75, 118
  on campus, 4
meaning making, 50, 58, 65–66, 112, 158, 161, 164
Meiklejohn, Alexander, 76
  Experimental College, 76
mentoring communities. *See* communities: mentoring
mentors, 170, 178
meritocracy, 89
metaphor vs. theme, 103
minority students. *See* students of color
mobility, 162, 164. *See also* upward mobility
multicultural education, 43, 48–49, 51–56, 59, 73–74
  approaches to, 51–54
  history of, 51
Multicultural Learning Community (MLC), 4, 91, 169
  capstone project, 105
  courses, 8, 96, 101, 102
  curricular goals, 97
  demographics, 98
  faculty, 128–29, 132, 136
  interviews, 4, 9, 10, 41, 84, 113–15, 119, 137, 139, 149
  methodology, 8–9, 83
  programmatic goals, 96
  study of, 4–6
  themes of, 7–9, 87–88, 101–6, 112–15, 139–40, 142, 150
  use of personal documents, 8
Myers Briggs Type Inventory, 90

name exercise, 104
narrative research, 7
  storytelling, 7, 133
Native Americans, 41, 66, 68, 98, 117
navigating, 30, 126
  academy, the, 149, 177, 181
  multiple worlds, 27
network, social capital, 22
nontraditional, 15, 51, 59, 76, 82, 167, 170
norms, middle-class. *See* expectations

opportunity, 13–14, 16, 21, 58, 59, 60–61, 63, 67, 70–71, 174–76, 182, 184
  gaps of, 16, 176
  orientation, 21

paradigm, 9, 51, 77, 139, 156, 177
  constructivist, 9
  shift, 51, 77
partnerships, 7, 49, 75, 89–90, 96, 136, 176, 179
  academic and student services, 3, 7
  between flagship institutions and school systems, 176
  intentional, 89, 96
part-time students, 48
pedagogical unsaid, 51
pedagogy, 5–7, 48–51, 53–59, 71–77, 88, 153
  critical, 7, 43, 49–51, 53–56, 59, 61, 71, 73–74, 88, 91, 103, 136
  definitions, 50
  emancipatory, 53–55
  multicultural, 7, 72, 75–77, 80, 85–86, 98, 108, 112, 118, 177
  reductionist, 54
peer groups, 30, 40, 83, 123, 126, 129–30, 132–33, 136–37, 152, 184
  cohorts, 5, 8, 10, 76, 78, 80, 98, 111, 115, 121, 129–31, 146, 164, 183
peers
  renegotiating relationships with, 43
  See also peer groups
Pell grants, 18
  recipients, 16
photomontage, 107, 108
political literacy, 53
Postsecondary Teaching and Learning, 100
  College English Transitions, 101
  Commanding English, 99
power
  differentials, 33, 39, 52, 61
  normative, 72
  relations, 50, 54
praxis, 7, 48, 51, 54–55, 74, 76, 144
  frameworks, 48
  integrated multicultural curriculum, 7
  learning-community design, 7, 73–74, 76, 84, 95–96, 118
programmatic support, 185
  interventions, 27
  policies, 27
  resources, 27
push-and-pull phenomenon, 23, 48, 175

race, 9–10, 14–16, 26–27, 38–39, 68–69, 71, 84, 132–33, 156–58, 164, 184
  racism, 16, 32, 40–41, 51, 62, 65, 67–69, 88, 102, 104, 107, 134, 154, 161
reflective writing, 5, 8, 66, 157
reframing, 185
remedial courses. *See* developmental courses
resilience, 4, 27, 33, 38, 177
roles, 3–4, 6, 21, 23–24, 26, 31, 36–38, 126–27, 180–83
  balance demands, 36
  competing life, 37
  juggling, 129
  multiple, 3, 24, 72

safe haven, 124. *See also* safe spaces
safe spaces, 70, 136, 141, 151, 184
Seattle Central Community College, 82, 83
self
  private, public and unknown, 107
self-authorship, 9, 65–66, 104, 139–42, 156, 161, 164, 175
  dimensions of, 66, 140, 142, 156, 161
  disequilibrium, 9, 62, 64–66, 108, 141–42, 149, 157
self-doubt, 26, 35
separation, 6, 41, 183
  shedding of one's social identity, 41
social advocacy, 162
social capital network, 22
social contract, 105, 112
social reconstructionist approach. *See* multicultural education
stereotypes, 52, 59, 133, 142–44, 146–47, 154
StrengthsQuest, 90
structural gate. *See* admissions
student-affairs professionals, 181
student-parent support groups, 182
students of color, 2, 9, 14, 16, 20, 25–26, 32, 39–40, 71, 74, 85, 97, 99, 114, 136
student-type learning communities. *See* learning communities
student voices, 4, 6–7, 9, 27, 50, 115, 118, 136, 142, 161

teaching communities. *See* communities: teaching
Temple University, 85
tenure process, 179
thematic categories, 9
time management, 37
Title III and IV students, 81
transition, 3, 5–6, 17, 30, 40, 43, 45, 90, 95–96, 100–101, 107
  from first to second year of college, 82, 84, 115, 183
  individuation, 43, 153
  surviving, 6, 45, 175, 183
TRIO SSS Programs, 3–4, 9, 14–15, 90, 96–97, 99, 100–101, 131, 182
  history of, 100–101
  purpose of, 100

underrepresented students, 3, 32, 186
underserved students, 48–50, 84, 129
universities
  accountability and fiscal restraint, 27
  flagship, 99, 168, 176, 185
  land-grant, 74, 99, 175–76, 178
  public and private, 19
  rankings and status, 49
  "university way," the, 34
University of Maryland, College Park, 85
University of Michigan, 85, 88
unwritten rules, 6. *See also* codebook
upward mobility, 16, 162, 168
  relationship between race and class, 16

validation of FG students' worlds, 35
Vygotsky, L. S., 77

Washington Center for Improving Quality Undergraduate Education, 82
ways of knowing, 43, 68, 97, 108, 132, 140, 142, 156, 158, 162, 173–74, 177
working class. *See* class: working
working while studying, 37
Wright, Frank Lloyd, 119, 138

Printed in the United States of America